AA003260

Sensors 2015

Topical Conference at the 2015 AIChE Annual Meeting

Salt Lake City, Utah, USA
8-13 November 2015

ISBN: 978-1-5108-1846-0

Printed from e-media with permission by:

Curran Associates, Inc.
57 Morehouse Lane
Red Hook, NY 12571

proceedings.com

Some format issues inherent in the e-media version may also appear in this print version.

Copyright© (2015) by AIChE
All rights reserved.

Printed by Curran Associates, Inc. (2016)

For permission requests, please contact AIChE
at the address below.

AIChE
120 Wall Street, FL 23
New York, NY 10005-4020

Phone: (800) 242-4363
Fax: (203) 775-5177

www.aiche.org

Additional copies of this publication are available from:

Curran Associates, Inc.
57 Morehouse Lane
Red Hook, NY 12571 USA
Phone: 845-758-0400
Fax: 845-758-2634
Email: curran@proceedings.com
Web: www.proceedings.com

TABLE OF CONTENTS

Functionalized Plasmonic Nanostructure Arrays for Direct and Accurate Mapping Extracellular pH of Living Cells Using SERS

Fang Sun, Shaoyi Jiang and Qiuming Yu*
Department of Chemical Engineering
University of Washington
2015 AIChE Annual Meeting
Salt Lake City

UNIVERSITY *of* WASHINGTON

Outline

- ❑ Introduction
- ❑ Quasi-3D Plasmonic Nano Arrays
- ❑ pH Detection using 4-Mercaptobenzoic Acid (4-MBA)
- ❑ Reproducibility and Interaction with Cations
- ❑ pH_e Distribution: Normal Cell vs. Cancer Cell
- ❑ Conclusions

1

Mapping extracellular pH of Living Cells

Significance

pH$_e$ Influences
- Cycle progression
- Proliferation
- Differentiation

Tumor Cell - Dysregulated pH
- Transformation
- Adhesion
- Migration
- Drug uptake

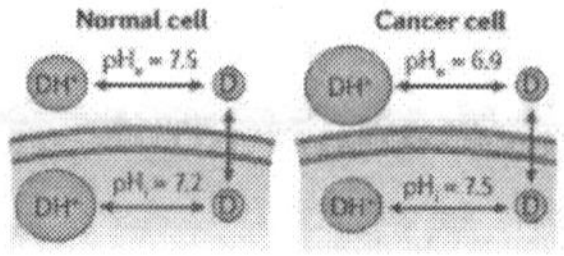

Conventional Methods
- pH microelectrodes
- MRI

Limitations
- Invasive method
- Low spatial resolution

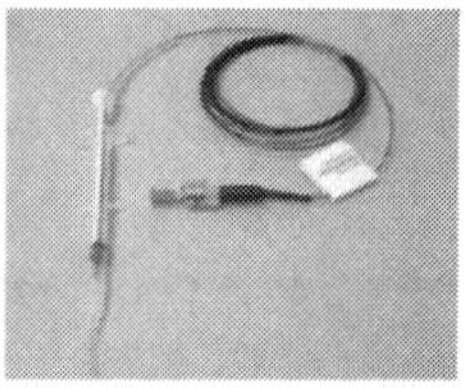

Microelectrode

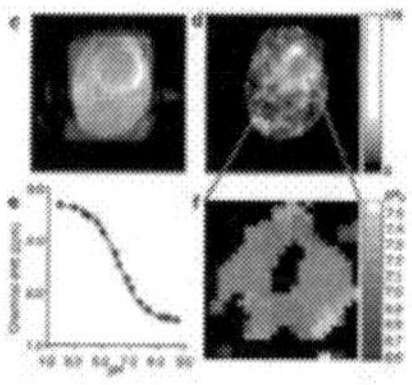

MRI

BA, Webb, et al. *Nature Reviews Cancer* . 2011, 9, 671-677

Surface Enhanced Raman Scattering

Enhancement Mechanisms:
- Electromagnetic enhancement (EF: 10^6-10^7)
- Chemical enhancement (EF: 10-10^2)

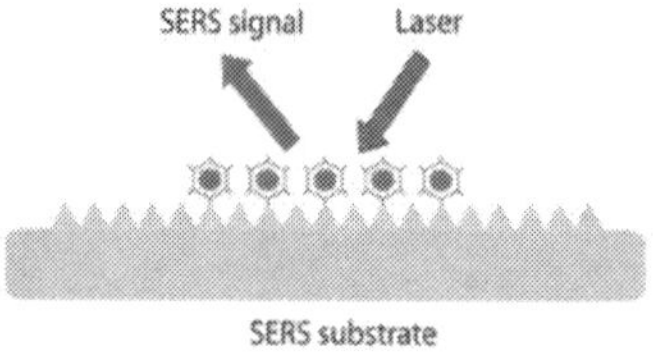

Advantages
- Ultra-sensitivity: single molecule detection
- Rich information: fingerprint spectra (molecular identity)

Applications

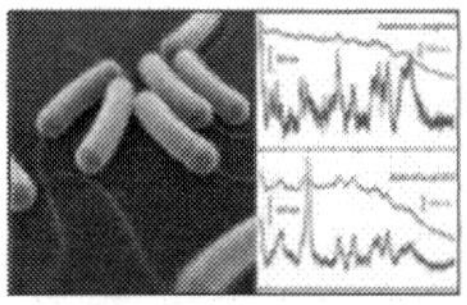

Bacteria

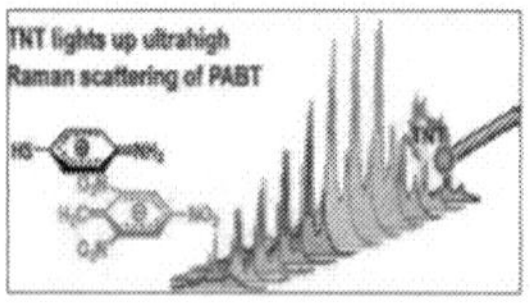

Chemical threats

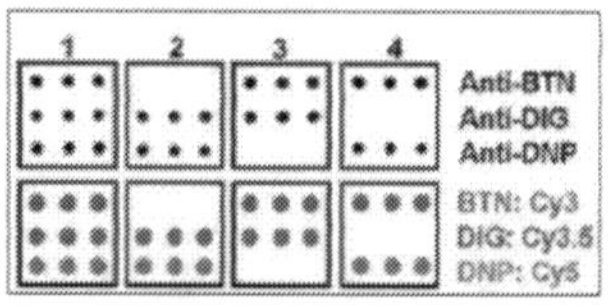

Diagnostics

J. Xu, et. al. *Small.* 2011, 3, 371
H. Zhou, et. al. *Anal. Chem.* 2011, 83, 6913
Y. Cao, et. al. *J.Am.Chem.Soc.* 2003, 125, 14676

SERS Substrate: Quasi-3D Plasmonic Nano Arrays

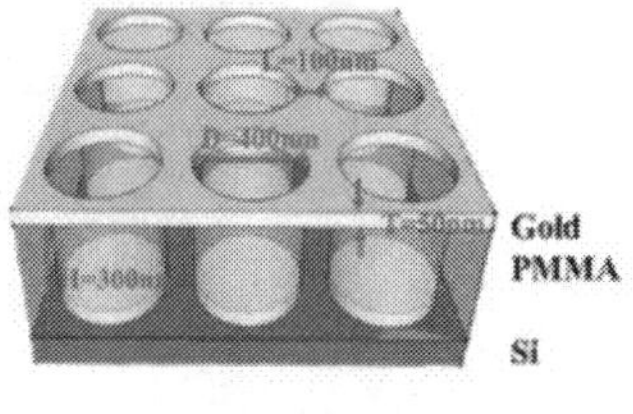

Q3D SERS substrate

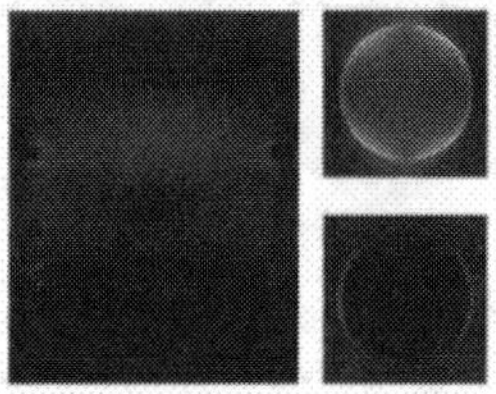

FDTD simulation

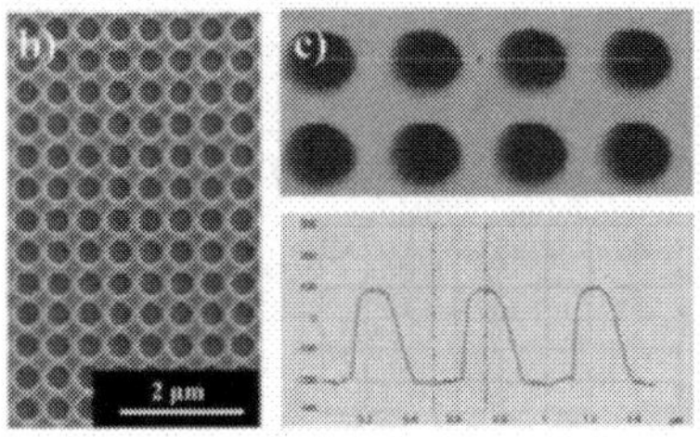

SEM and AFM

- ✓ High sensitivity
- ✓ High reproducibility

F. Sun, et al. *Anal. Chem.* 2014, 86, 2387-2394

pH Detection using 4-Mercaptobenzoic Acid (4-MBA)

4-MBA modified SERS substrate

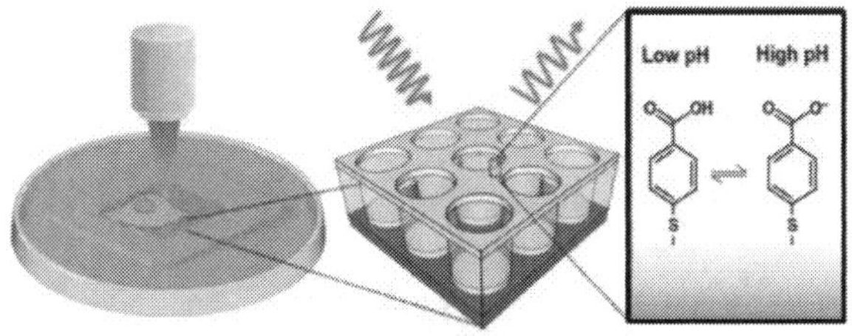

SERS spectra with increasing pH

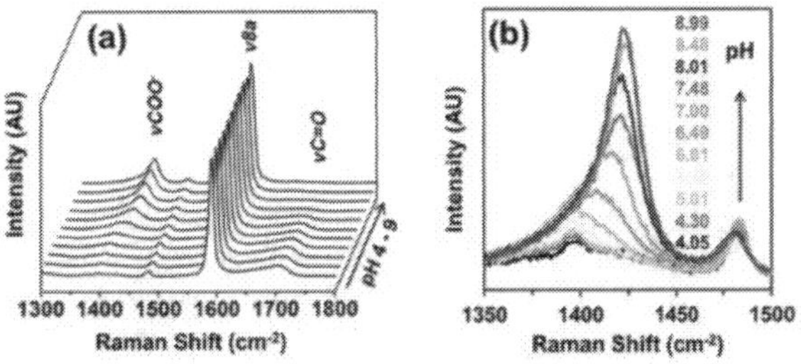

- ✓ pH detection is achieved by monitoring the SERS spectral changes of the peak at 1420 cm^{-1} ($vCOO^{-}$) of 4-MBA.

F. Sun, et al. *Biosensors and Bioelectronics.* 2015, 73, 202-207

Reproducibility and Interaction with Cations

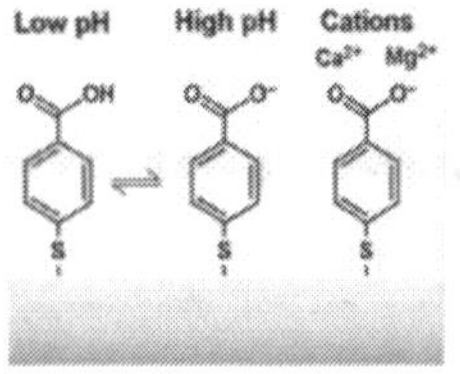

SERS contour of 4-MBA from 32 spots

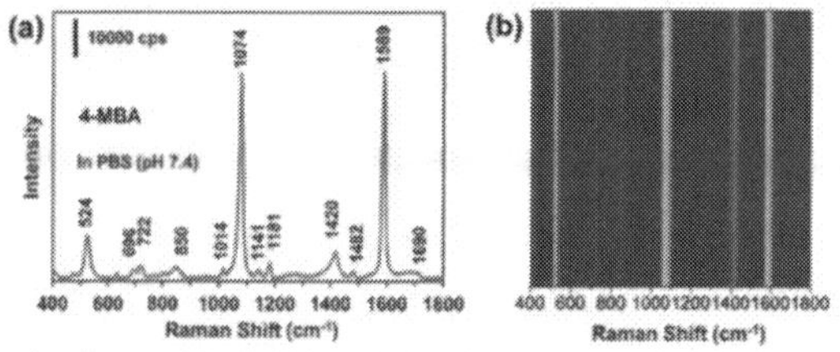

Influences from the cations and calibrations

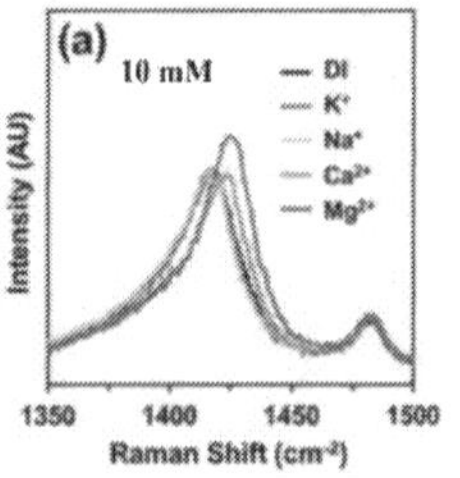

- Intracellular mimic (IC)
- Cell culture medium (CM)
- PBS
- DI water
- Phosphate citrate (PC)

✓ Reproducibility
✓ Reliability

F. Sun, et al. *Biosensors and Bioelectronics*. 2015, 73, 202-207

pH_e Distribution: Normal Cell vs. Cancer Cell

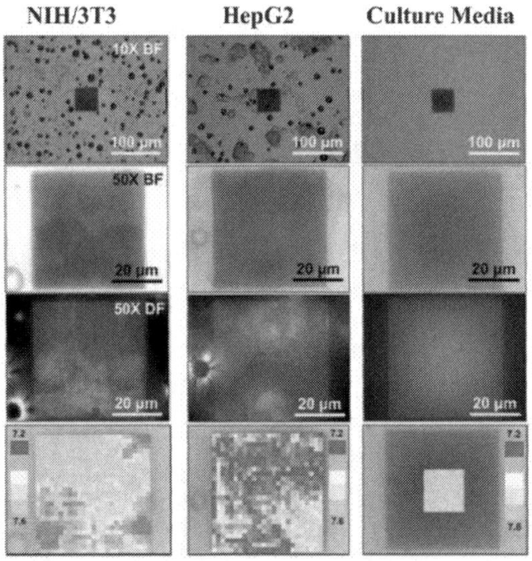

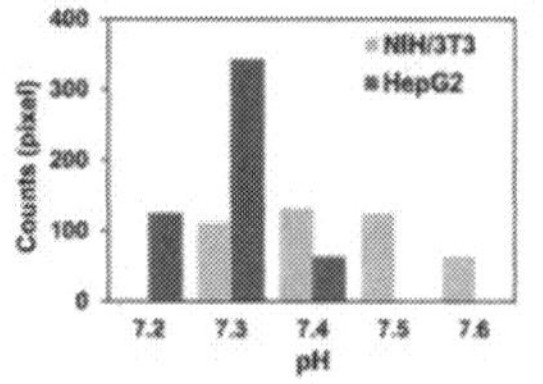

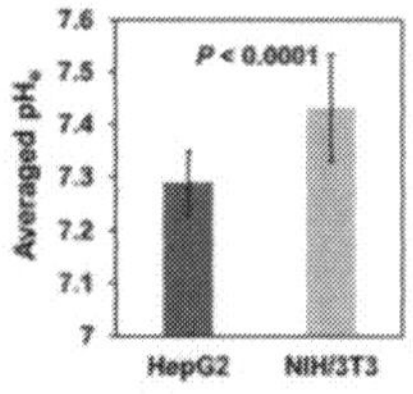

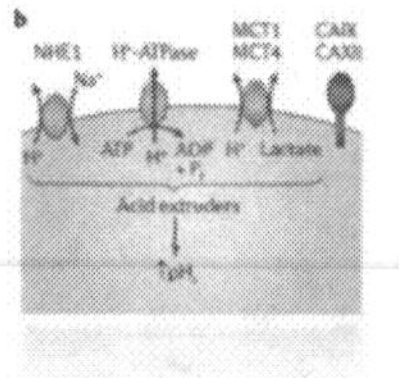

✓ pH_e domains are observed
✓ Tumor cells show more acid averaged pH_e

F. Sun, et al. *Biosensors and Bioelectronics*. 2015, 73, 202-207

Conclusions

- Q3D-PNAs SERS substrate demonstrated excellent reproducibility.
- 4-MBA modified Q3D-PNAs SERS substrate can detect the pH variation. Resolution: 0.1 pH unit.
- Divalent cations influenced the pH detection by 4-MBA.
- pH_e distributions of normal and tumor cells were mapped and imaged.
- Tumor cells demonstrated more acid averaged pH_e.

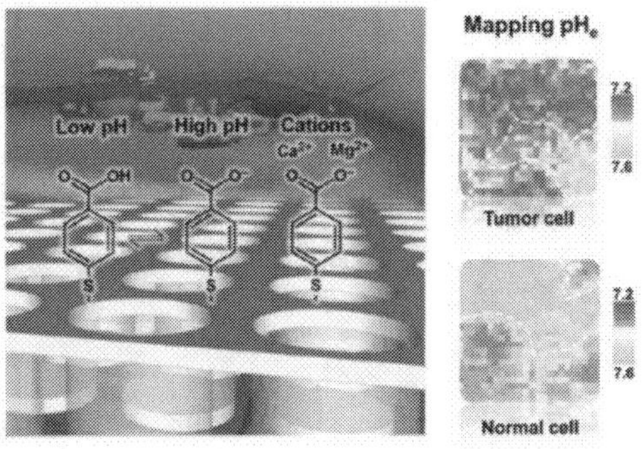

F. Sun, et al. *Anal. Chem.* 2014, 86, 2387-2394
F. Sun, et al. *Biosensors and Bioelectronics.* 2015, 73, 202-207

8

Acknowledgements

Prof. Shaoyi Jiang Research Group

Prof. Qiuming Yu Research Group

Funding Sources

DARPA NSF NNIN Nanoscale Science Engineering & Technology

9

Questions?
ESPRESSO

416957 Conductivity, pH, and Oxidation Reduction Potential Sensing with a Simple Three-Electrode Microfabricated System

Abstract:

Detection of pH and oxidation reduction potential (ORP) in aqueous solutions typically requires a reference electrode that supplies a constant and known voltage regardless of solution conditions. These electrodes are challenging to construct for benchtop operation and are even more difficult to microfabricate. Current microfabricated reference electrodes and ion-selective electrodes use relatively difficult processing steps and have relatively short user lifetimes. For instance, silver/silver-chloride devices require construction of a constant concentration silver-chloride reservoir, and ISFET fabrication requires silicon doping.

We have constructed a three-electrode platinum sensor that can detect the conductivity, pH, and ORP of an aqueous solution. The three-electrodes system measures conductivity and ORP sequentially using an AC and a DC signal, respectively, passed through two electrodes. To measure pH, a controlled current signal generates both a constant voltage and a pH-varying voltage on the three-electrode system. This novel method, unlike other micro-fabricated reference electrodes, uses platinum electrodes and, hence, has an almost unlimited lifetime. Further, the fabrication procedure is comparatively easy and cheap: only single layer of platinum deposition on a glass wafer was necessary. The novel pH and ORP sensor can measure conductivity, pH, and ORP in non-oxidizing (below 1 ppm chlorite) solutions. Measurements in oxidizing solutions (above 12 ppm chlorite) are also possible although modification of one electrode with a surface coating is necessary.

Copyright © American Institute of Chemical Engineers. All rights reserved.

435200 A Low-Noise Glass Nanopore: Comparison of Nanopores in Glass and Silicon Nitride

Abstract:

Synthetic nanopores provide benefits over biological nanopores in sensing applications; they are stable over a broader range of environmental conditions and offer greater flexibility with regard to pore size and surface chemistry.[1] Micromachining processes have been well developed over the past five decades with the growth of IC technology, which can be employed to fabricate synthetic nanopores. For example, nanopores may be milled straightforwardly in thin, solid-state membranes using a focused ion or electron beam. Since nanopore-based sensing devices rely on changes in conductance as analytes block the nanopore or are translocated through it, there have been many attempts to reduce signal noise to improve sensitivity. For example, insulating material (*e.g.*, PDMS, silicon dioxide or silicon nitride) is commonly deposited on the membrane surface to decrease noise. Although most micromachined nanopores are silicon based, it has been reported that glass capillary nanopores show lower noise and a higher signal-to-noise ratio for DNA translocation compared to nanopores in silicon nitride membranes.[2] However, the pulled capillary glass nanopore exhibits a smaller current due to its long conical shape and is not amenable to high-throughput manufacturing. Recently, a low-noise, solid-state nanopore based on an insulating substrate has been introduced in which a thin silicon nitride membrane is attached to a quartz substrate to decrease low frequency noise by several orders of magnitude.[3]

In this work, the creation and characterization of a glass nanopore is presented. Both the membrane and the substrate are derived from a single glass wafer using batch micromachining. By thinning the substrate with a dilute mixture of hydrofluoric and hydrochloric acid,[4] submicron-thick membranes are successfully produced. In addition, a comparison of the glass nanopore versus the silicon-based nanopore is investigated based on the stability of the base current, the measured noise-level (*e.g.*, RMS ionic current noise and current power spectral density in frequency domain), and the presence of undesired peaks upon changing applied voltage polarity. A relation between noise level and pore sizes is also established. To test the nanopore as a sensor, the glass chip with membrane and nanopore was sandwiched between two buffer-filled chambers. Carboxylate-terminated polystyrene beads were placed in one of the buffer-filled chambers, and a constant electric field was applied across the pore. Due to its negative charge, the bead migrates to the pore mouth and blocks it giving rise to a decrease in conductance. Using this test, the performance of the nanopores in glass and those in silicon nitride are compared.

[1] Dekker C. Solid-state nanopores. Nat. Nanotechnol. 2007; 2: 209-215.

[2] Steinbock LJ, Bulushev RD, Krishnan S, Raillon C, Radenovic A. DNA Translocation through Low-Noise Glass Nanopores. ACS Nano. 2013; 12: 11255-11262.

[3] Lee M-H, Kumar A, Park K-B, Cho S-Y, Kim H-M, Lim M-C, Kim Y-R, Kim K-B. A Low-Noise Solid-State Nanopore Platform Based on a Highly Insulating Substrate. Sci. Rep. 2014; 4: 1-7.

[4] Iliescu C, Chen B, Miao J. On the wet etching of Pyrex glass. Sens. Act. A. 2008; 143: 154-161.

Copyright © American Institute of Chemical Engineers. All rights reserved.

430572 Detection of Acetone in Exhaled Breath By Gold Nanoparticle Gas Sensors

Abstract:

Detection of acetone in exhaled breath by gold nanoparticle gas sensors

Zhenzhen Xie[a], Xiao-an Fu[a*]

a Department of Chemical Engineering, University of Louisville, Louisville, KY 40292

*Corresponding author: xiaoan.fu@louisville.edu

Abstract

With the worldwide prevalence of diabetes, the methods for diagnosing have received considerable interest. The conventional method of blood tests is painful and has some limitations. Acetone in the human breath has been identified as an important biomarker for diagnosis of diabetes. The average concentration of exhaled acetone in the healthy human breath is found to be lower than 0.8 ppm, while that is higher than 1.8 ppm in diabetic patients. Gas sensors with sub-ppm acetone detection capacity play a significant role in the development of noninvasive monitors for diabetic patients. Gold nanoparticle-based sensors offer several advantages over conventional metal oxide-based sensors, including low power consumption, flexibility, ambient temperature operation and good chemical selectivity. They can be operated at room temperature and have shown limits of detection in the parts-per-million (ppm) or even parts-per-billion (ppb) level.

The main objective of this work is to increase the sensitivity of the sensors between different gold-thiolate nanoparticles for the detection of volatile organic compounds (VOCs) in human breath and dry air, and also the effect of the relative humidity. Gold nanoparticles with small size (< 5 nm) were synthesized. We found that the response follows a linear regression with increasing concentration of VOC in the range from 0.1 ppb to 5 ppm. Our results show that gas sensors based on gold nanoparticles could be used to distinguish healthy human from diabetic patients.

Keywords: diabetes, acetone, gold nanoparticle, sensors

Copyright © American Institute of Chemical Engineers. All rights reserved.

429421 Carbon Nanofiber Mats for Surface Enhanced Raman Spectroscopy

Abstract:

Carbon Nanofiber Mats for Surface Enhanced Raman Spectroscopy

by

Juan F. Yee-De-León[1], Victor H. Perez-Gonzalez[1], Roberto C. Gallo-Villanueva[1], Sergio O. Martinez-Chapa[1,*]

[1] Sensors and Devices Research Group, School of Engineering and Sciences, Tecnológico de Monterrey, México

*Correspondence should be addressed to smart@itesm.mx

Abstract

Surface Enhanced Raman Spectroscopy (SERS) is a powerful readout technique that yields qualitative and quantitative information of chemical species and biomolecules at low concentrations. The effectiveness of the technique is closely related to the SERS-active substrate employed. Currently, nano-fabrication techniques have enabled the development of highly active and reproducible SERS-substrates, however, very specialized personnel and equipment are needed to construct these substrates, increasing the related cost of sample analysis using SERS as the readout technique. This situation has hampered the use of SERS in real life situations, which has motivated to the quest of novel nanomaterials and methodologies for producing homogenous, highly sensitive, scalable and low cost SERS-active substrates for chemical trace detection and biological applications.

In this work, polyacrylonitrile (PAN) nanofiber mats were fabricated via far-field electrospinning. These mats were subjected to a carbonization step to obtain nanofiber mats composed of carbon. The carbon nanofiber mats obtained were oxidized using a solution composed of 3:1 (H_2SO_4:HNO_3) v/v, and later, the surface of the nanofibers was functionalized with aminosilanes, in order to attach gold nanoparticles to the carbon nanofiber mats. Here, we present a first approximation to the use of carbon nanofiber mats as structural three-dimensional scaffolds for supporting metallic nanoparticles for its use as SERS-active substrates. The structure's high surface area and porosity, derivatives from the three-dimensional entanglement of the nanofibers, allows easy access and interaction of more analyte molecules with the metallic nanostructures located onto the nanofiber's surface. A higher number of metallic nanoparticles adsorbed increases the sensitivity and saturation levels of these substrates with regard to 2D SERS-active substrates.

Copyright © American Institute of Chemical Engineers. All rights reserved.

430056 A Comparison of Cobalt and Gold Functionalized TiO2 Nanotubes Based Sensing Platform for Enhanced Electrochemical Detection of Volatile Organic Biomarkers

Abstract:

One class of biomarkers that is known to have associations with certain diseases, and yet has found limited use as a diagnostic tool is volatile organic biomarkers (VOBs). From a healthcare standpoint, sensing and detection of VOBs from the breath is important for early diagnosis of several pulmonary diseases, including tuberculosis, whose distinct biomarkers have been successfully identified by researchers using mass spectroscopy. Time of detection, cost, equipment portability, and performance have been identified as some of the major roadblocks in the development of a suitable device for point-of-care (POC) detection of tuberculosis (TB). In this study, we demonstrate the use of metal functionalized TiO_2 nanotubes array (TNA) based robust sensing platform for rapid electrochemical detection of the four prominent TB biomarkers viz. methyl nicotinate, methyl p-anisate, methyl Phenylacetate, and o-phenylanisole.

Solid-state TiO_2 nanotube array has demonstrated to be a promising sensing platform for VOCs detection. Highly ordered TNA were synthesized through electrochemical anodization, and functionalized separately with cobalt and gold, which were determined as leading frontrunners for efficient binding with the VOBs. Cobalt functionalization was first achieved using the incipient wetting impregnation method (aka dip coating method) by sonicating the TNA in 0.1M $CoCl_2$ solution for 30 minutes. The sensor response was recorded through electrochemical detection of the four VOBs based on the observable change in current at fixed low bias upon exposure of the sensor to the vapors of the biomarker. In an effort to increase the sensitivity of the sensor, cobalt was incorporated into the titania nanotube lattice through a one step in-situ anodization-co-functionalization process in a CoF_2 based ethylene glycol solution. As expected, enhanced sensor response was achieved with reduced time of detection. Further, TNA was functionalized with gold using an electroless deposition technique. Results indicate that sensor response from Au-TNA increased manifold over its Co-TNA counterparts.

The sensor substrates were characterized using SEM, TEM, and XPS. Indirect band gap was calculated based on UV-Vis absorbance. Further, a mechanism has been proposed to describe the attachment of predominate biomarkers to the sensor surface, based on band theory. Overall, the robust synthesis and approach of the sensor demonstrates promise for developing and commercializing a hand-held device for point-of-care diagnosis of pulmonary diseases.

Copyright © American Institute of Chemical Engineers. All rights reserved.

430207 Biocompatibility Evaluation of Skin Wearable Silver Nanowire Biosensor Systems

Abstract:

Recent advances *in vitro* diagnostics on human skin or organ surfaces rely on the use of sensors. Wearable diagnostic and therapeutic devices are ubiquitous to health care, where technology can improve patient's life. Due to the high aspect ratio and excellent electrical conductivity, silver nanowires (AgNWs) have been widely explored for flexible/stretchable conductors that can accommodate large deformations, make intimate contact with curvilinear surfaces (such as human skin) and maintain high conductivity under large strain. A plethora of wearable devices using AgNWs have been demonstrated in the past few years, such as strain (motion) sensors, pressure sensors, touch sensors, and electrophysiological electrodes. However, biocompatibility of silver nanowires to human skin is largely unknown, which is critical for these wearable devices as many of them are in contact with skin for long time for long-term, continuous health and wellness monitoring. Lack of biocompatibility in a skin-attachable sensor, can result in complications including: cell cytotoxicity, skin irritation, skin sensitization, skin chronic inflammation and dermatitis.

We fabricated a highly conductive and stretchable conductor with AgNWs embedded in the surface layer of poly(dimethylsiloxane) (PDMS). Assessment of the materials biocompatibilities on skin is currently performed with the use of animals. Alternative non-animal related *in vitro* methods are demanded by Animal welfare and EU legislation. We have developed novel *in vitro* biocompatibility test methodology for screening of skin-wearable AgNW/PDMS sensors. These silver sensor components were assessed *in vitro* using two skin cell lines and 3-dimensional tissues. Three characterized in vitro assays were used to measure the cytotoxicity, irritation and sensitization of the skin-attachable sensor materials. These *in vitro* assays could meet the FDA requirement, as well as minimize the usage of live animals. We also characterized these AgNW/PDMS sensors to ensure that their electrical conductivities were performed functional for different applications, such as ECG and hydration sensors, without eliciting any undesirable local or systemic effects, including cytotoxicity and skin irritation.

Copyright © American Institute of Chemical Engineers. All rights reserved.

438021 High Throughput Mobile Phase Microfluidic Sensing and Analytical Platforms

Abstract:

Miniaturization and microfluidics has enabled a new class of instruments capable of integrated multistep processes. Such micro-total analysis systems are well suited for a wide range of cellular and molecular sensing applications, that can be conducted at the point of care or point of need. A critical bottleneck to full microscale integration has been a lack of native on-chip particle/cell concentration and alignment technologies. Inertial focusing provides a solution to each of these hurdles by passively focusing particles to predictable, precise lateral and longitudinal positions. The enabling fluid dynamics of inertial focusing, as well as the possibilities and limits for the technology, will be presented. Subsequently, applications for inertial focusing, including sample preparation, cellular processing and optical analysis, will each be discussed in turn.

Copyright © American Institute of Chemical Engineers. All rights reserved.

437463 Biosensors for Genes, Pathogens, Parasites, Biomarkers and Toxins

Abstract:

Our work over the last decade has examined a number of platforms for detecting significant biological entities. This includes classical fluorescence based measurements (such as intracellular NAD(H) in bacterial and cell cultures), optical evanescent field sensors for proteins, magneto-elastic film devices and electromechanical resonators, especially cantilevers that have integrated piezoelectric films. In this talk, I will focus on the latter device as its design has yielded extraordinary sensitivity. The methods we have developed allow for eliminating false negatives, a critical performance requirement for bioterrorism, medical, environmental and food safety applications.

The cantilever sensors are self-excited devices that exhibit high-order modes at ~ 0.1 to1 MHz and show sub-femtogram sensitivity. One significant property they demonstrate is that nonspecific binding is low or absent due to the surface being under constant out of plane oscillation. Several examples of practical importance (*E. coli* O157:H7, biomarkers, waterborne parasites, food and water toxins, and *B. anthracis*) will be illustrated using both antibody-based sensors and specific gene sequence as a molecular identifiers, and without an amplification step.

Acknowledgement. The author is thankful for the following grants that made the biosensor research possible. NSF-CBET 0828987. NSF-CBET 1159841. EPA STAR R833007. EPA STAR R833829. USDA 2006-51110-03641. DoT-PA-26-0017-00. NBIB 5R01EB000720. NCI R43GM083595-01. PA DoH GRID-2011-5A

Copyright © American Institute of Chemical Engineers. All rights reserved.

437249 Portable Sensors Based on Integrated Photonics

Abstract:

Innovation in technology routinely leads the way for discovery in chemistry and biology. Most notably, x-ray diffraction data was instrumental in the elucidation of the structure of DNA. To explore the inherent complexity present in biological systems, existing technologies are being pushed to their limits. Once again, scientists are looking to engineers to create innovative solutions to enable their exploration and discovery. However, the emphasis is governed by the specific application. While sensitivity enhancement is critical to achieve single molecule detection levels, in many applications, improved stability and reduced false-positive rates are of higher importance. Therefore, specific challenges posed by discussions with medical researchers and physicians should also be considered. For example, we have recently demonstrated a fully integrated, portable polarimetric fiber sensor instrument for characterizing the mechanical properties of visco-elastic materials. By combining polarization-maintaining fiber, an integrated laser and a polarimeter, all free-space components which were required in previous systems have been eliminated. This reduction in complexity eliminates alignment steps and decreases the overall footprint of the system. A side-by-side comparison with a loadframe shows similar accuracy when determining a series of visco-elastic materials with Young's Modulus values spanning over an order of magnitude. Moreover, the system's performance in a range of environments, ranging from in a fumehood to on an optical table, shows sufficient robustness against environmental noise to enable characterization of low Young's Moduli materials, such as tissue. This portable system shows promise for rapid testing and characterization of human tissue samples. This will enable numerous types of biomedical research investigations and new diagnostic methods.

Copyright © American Institute of Chemical Engineers. All rights reserved.

436844 Hydrogel Modules for Sensitive Bioassays

Abstract:

Due to their hydrophilic, biocompatible, and highly tunable nature, hydrogel materials have attracted strong interest in the recent years for numerous biotechnological applications. In particular, their solution-like environment and non-fouling nature in complex biological samples render hydrogels as ideal substrates for biosensing applications. We will discuss newmicrofabrication techniques for synthesizing encoded particles and microfluidic devices which make use of hydrogel materials for bioassays. We will highlight our recent work on microRNAs which have emerged as potential biomarkers for several diseases including cancer, diabetes, and cardiovascular disease, due to their unique dysregulation patterns under disease conditions. We will also discuss methods to encode and decode particles via shape.

Copyright © American Institute of Chemical Engineers. All rights reserved.

428945 Remote Manipulation with Microlablets and Control of Enzymatic Reaction

Abstract:

Progress in microfabrication technologies makes it possible to design and fabricate autonomous self-assembling microparticles called lablets (under 100 mm in size), which can implement programmable microscale electronic chemistry, forming a bridge between electronic and chemical computing. Due to limited energy storage capacity of the lablets and high material density, their movement in aqueous environment over larger distances and attachment to a predefined substrate (docking station) remains a challenge. The aim is precise positioning of the lablet above the docking station in aqueous environment, followed by controlled attachment of the lablet on pre-defined part of the docking station. The position of the lablet with respect to the docking station has to be within the range of 10's of micrometers.

The lablets are mainly composed of silicon (density 2.57 g.cm^{-3}), hence they are denser than water and thus remote manipulation with the lablets above the docking station in aqueous environment is not possible due to their sedimentation. 150 - 200 mm small surfactant stabilized oil droplet (kerosene, density 0.80 g.cm^{-3}) were be attached to the lablets in order to decrease their overall density. The oil microdroplets with attached lablets were observed to be floating on the water-air interface. Since one side of the lablets surface is formed from silicon (Si) while the other one contains also gold electrodes (for communication with the dock station), it is required to attach the lablets to the oil droplets in such a way that the gold electrodes will be immersed only in the water phase and thus be able to communicate with the dock station. The immersion of the lablets in the oil phase can be changed by hydrophilic/hydrophobic modification of the lablet's surface.

In order to make the remote manipulation possible, magnetic (iron-oxide) nanoparticles with a hydrophobic modification (oleic acid stabilized) were added to the oil (kerosene) phase to create a ferrofluid. The ferrofluid will consequently move in a magnetic field and carry the lablet with it. The magnetic field is created by four electromagnets (the distance between two opposite electromagnets will be 25 mm), allowing us to precisely manipulate with the lablets in the X,Y (planar) direction. Depending on the voltage applied and the concentration of magnetic particles in the ferrofluid, lablets can be moved with a velocity of 10'-100's mm/s. The attachment of the lablet to the docking station will be achieved by placing another electromagnet bellow the docking station, hence we will be able to move/elongate the ferrofluid microdroplets in the Z (vertical) direction that will consequently lead to the connection of the lablet with the docking station.

The lablets can also act as carriers of active substances, i.e. liposomes with encapsulated substrate attached on the lablets surface. The release of the encapsulated content can be triggered by i) radiofrequency heating which causes a reversible phase transition of the lipid bilayer; or by ii) electroporation of the lipid bilayer. By controlled release of the substrate, we are able to trigger an enzymatic reaction on-demand. This principle makes it possible to deliver unstable or reactive active ingredients that cannot be formulated into traditional dosage forms.

Copyright © American Institute of Chemical Engineers. All rights reserved.

435312 Enzyme Immobilization By Polydimethylsiloxane Stamping for Glucose Sensor Fabrication

Abstract:

Abstract

Enzyme Immobilization by Polydimethylsiloxane Stamping for Glucose Sensor Fabrication

Bo Wang, Bonhye Koo, Harold G. Monbouquette*

Chemical and Biomolecular Engineering Department, University of California, Los Angeles, CA 90095, USA

Most glucose sensors are based on electrochemical detection of hydrogen peroxide (H_2O_2) generated from the glucose oxidase (GOx)-catalyzed oxidation of glucose, and immobilization of the enzyme is a critical step in biosensor fabrication. Here, a new and effective method for GOx immobilization is presented. The enzyme is successfully transferred from a polydimethylsiloxane (PDMS) stamp surface onto a chitosan-coated electrode. Compared to manual enzyme immobilization methods, the enzyme layer generated by PDMS stamping is thinner, which leads to fast sensor response and high sensitivity. The sensor construction procedure begins with the coating of polypyrrole and Nafion on a platinum (Pt) electrode surface sequentially as permselective polymers to exclude anionic and cationic interferents (*e.g.*, ascorbic acid and dopamine).[1] A chitosan film is electrochemically deposited on top of the Nafion to serve as a suitable adherent layer for enzyme immobilization.[2] GOx and bovine serum albumin (BSA) are transferred together onto the chitosan surface using a PDMS stamp. Then, the stamped layer is exposed to glutaraldehyde vapor for cross-linking. The layering process of biosensor fabrication is monitored by electrochemical impedance spectroscopy (EIS) and the chitosan surface before and after enzyme stamping is characterized by scanning electron microscopy (SEM). In testing with varying concentrations of glucose and interferents, the sensors show high sensitivity, low detection limit, good selectivity, and fast response time. This enzyme immobilization method can be used in principle to immobilize various enzymes for creation of biosensors for applications both *in vitro* and *in vivo*.

Reference

1. Tseng, T.T.C. and H.G. Monbouquette, *Implantable microprobe with arrayed microsensors for combined amperometric monitoring of the neurotransmitters, glutamate and dopamine.* Journal of Electroanalytical Chemistry, 2012. **682**: p. 141-146.

2. Tseng, T.T.C., J. Yao, and W.C. Chan, *Selective enzyme immobilization on arrayed microelectrodes for the application of sensing neurotransmitters.* Biochemical Engineering Journal, 2013. **78**: p. 146-153.

Copyright © American Institute of Chemical Engineers. All rights reserved.

428540 Using the Personal Glucose Meter to Quantify a Wide Range of Targets

Monday, November 9, 2015: 3:53 PM
Canyon C (Hilton Salt Lake City Center)

Tian Lan, Research Park, University of Illinois at Urbana-Champaign, Champaign, IL, Yu Xiang, Chemistry, TsingHua University, Beijing, China and Yi Lu, Chemistry, University of Illinois at Urbana-Champaign, Urbana, IL

Using the personal glucose meter for convenient quantification of wide range of targets

Tian Lan, Yu Xiang and Yi LU

Abstract:

Convenient, low-cost and quantitative detection of broad range of targets has the great potential to revolutionize today's practice for medical and environmental monitoring. Despite years of research, very few such devices have been commercialized at a large scale. Today's personal glucose meter (PGM) is the culmination of decades of R&D and commercialization efforts aimed to satisfy the daily need of millions of diabetes patients worldwide to monitor their blood glucose levels. The PGM is the almost the ideal sensing platform, which offers convenience, low-cost and quantitative, except it only quantifies one target, glucose.

Instead of designing a new platform that can be time consuming and costly, we have developed an innovative method that adapts the widely available PGM for quantifying a wide range of targets, including metal ions, small molecules, proteins and nucleic acid sequences. An enzyme invertase, which hydrolyzes sucrose into glucose, is conjugated to recognition molecules, such as aptamers, DNAzymes, antibodies and DNA, and relays the non-glucose target binding events to the presence and concentration of glucose, which can be measured using PGM. Using this approach, we have demonstrated quantification of metal ions using DNAzymes, small molecules using aptamers and antibodies, various protein markers using antibodies, and viral DNA fragments using complementary DNA as shown in Table-1.

Table 1. New targets detected and quantified by BGM using our novel technology

Type	Example	Detection limit
Diabetic biomarkers	HbA1C	3 mg/L
Disease biomarkers	INF-γ (a TB marker)	2.8 nM
Cancers	Prostate cancer antigen (PSA)	0.4 ng/mL
Small organic molecules	Cocaine adenosine	5.3 μM
Organic toxins	Ochratoxin A	6.8 ng/mL
Metal ions	Pb and U	9.8 nM
Viral DNA	Hepatitis B	40 pM

In conclusion, we have demonstrated an innovative way of leveraging the existing PGM hardware for quantification of non-glucose target using different recognition molecules. The technology can be used to develop new sensing platform using the technologies developed for the PGM over the last decades.

430671 Sequence-Specific Nucleic Acid Detection Based on Blockade of a Nanopore in a Thin Glass Membrane

Abstract:

A new platform for sequence-specific nucleic acid (NA) detection with binary response has been demonstrated. The detection of nucleic acid sequences is useful for diagnosing the presence of bacteria, especially those proven to be harmful to public health. A platform for low cost, reliable detection of the presence of pathogenic bacteria has a wide variety of applications, ranging from the diagnosis of bacterial infections to identification of contaminated food or water. Most existing methods are complex and require special reagents, as they rely on polymerase chain reaction (PCR) for target sequence amplification and fluorescence or enzyme labels. In our previous work, we described our first-generation, PCR-free, label-free system based on polystyrene beads conjugated with uncharged peptide nucleic acid (PNA) as sequence-specific probes, and demonstrated a detection limit of 10 fM.[1,2] A drawn glass pipette tip served as the micropore and was placed between two buffer-filled chambers, one of which contained the bead-PNA conjugates. Since the bead-PNA conjugates carry little to no charge, they do not respond to an electric field applied across the pore. However, in the presence of target NA sequence, the bead-PNA conjugates acquire enough negative charge to become mobile by hybridizing the NA sequence of interest. Upon pore blockade by the beads, a change in the pore conductance occurs, which is reflected in a step-decrease in current that signals the presence of the target nucleic acid.[1,2]

At lower concentrations of NA where fewer target NA molecules hybridize per bead, smaller beads are expected to have a higher mobility. Accordingly, by decreasing both the size of the pore and the beads, detection limits below 10 fM may be achieved. Motivated by this analysis, we have fabricated thin (1 μm or thinner) glass membranes in which a nanopore is milled with a focused ion beam (FIB). This device has two benefits: 1) By milling a pore via FIB, the pore size can be reduced significantly compared to drawn glass pipette tips, which lie in the micrometer range; 2) Since the nanopore is milled in a planar substrate, it can be mass-manufactured and can be assembled into devices more easily. To date, the nanomachined devices have proven capable of distinguishing between complementary and non-complementary sequences of ssDNA at various bead sizes. In addition, a promising trend has been observed showing that with decreasing bead size, higher mobility is attained at lower NA concentrations. These results support the expectations that detection limits less than 10 fM may be achieved with the nanomachined device and that this work could give rise to a diagnostic platform for the rapid, sensitive, and cost-effective detection of NA of specific sequence.

[1] Esfandiari L, Monbouquette HG, Schmidt JJ. Sequence-specific Nucleic Acid Detection from Binary Pore Conductance measurement. J. Am. Chem. Soc. 2012; 134: 15880-15886.

[2] Esfandiari L, Lorenzini M, Kocharyan G, Monbouquette HG, Schmidt JJ. Sequence-Specific DNA Detection at 10fM by Electromechanical Signal Transduction. Anal. Chem. 2014; 86: 9638-9643.

Copyright © American Institute of Chemical Engineers. All rights reserved.

429352 Quantitative Analysis of Multivalent Protein Binding Using a Novel Combinatorial Lipid Array

Abstract:

Nolan Worstell[1], Joshua Weatherston[1], Pratik Krishnan[1], Ishan Bajaj[1], M.M. Faruque Hasan[1], Hung-Jen Wu[1]

[1]Dept. Chemical Engineering, Texas A&M University, USA

Most biochemical processes in the intracellular environment begin by membrane recruitment, protein transport from 3-D space to a 2-D membrane surface. This is achieved by multivalent binding interactions that synergistically combine weak affinity monovalent lipid-protein binding events to increase binding affinity and specificity. These multivalent binding events are essential for protein recognition, but the mechanism is poorly understood. To better understand the multivalent binding mechanism, quantitative analysis of multivalent membrane recruitment onto the cellular membrane is essential. We have developed a nanocube sensor coupled with complex reaction analysis to quantitatively explore the multivalent binding mechanism. The nanocube sensor is surrounded by a lipid bilayer that possesses the same physical and chemical properties as cell membranes. This biomimetic surface then enables the label-free detection of multivalent binding interactions by observation of the absorption spectra shift of the localized surface plasmon resonance (LSPR) peak. This biosensor works with a standard laboratory plate reader for high-throughput binding kinetic analysis. The simple protocol ("mix-and-then-detect") allows any end users to perform the analysis in their own laboratories. We have successfully explored many essential multivalent binding events, including Cholera toxin-glycolipid and Shiga toxin-ganglioside interactions. Moreover, we have introduced a complex reaction analysis technique to analyze the rich kinetic data acquired by the high-throughput lipid array and model the binding cooperativity among multivalent binding proteins, ligands, and accessory molecules. This technique can also be used to study other biomolecule adhesion, including nucleic acid and pathogen adhesions.

Copyright © American Institute of Chemical Engineers. All rights reserved.

410635 Real-Time Detection of Lipid Bilayer Assembly and Detergent-Initiated Solubilization

Abstract:

The cellular membrane governs numerous fundamental biological processes, and therefore, developing a comprehensive understanding of its structure and function is critical. However, because of its inherent complexity, this challenge is as yet unsolved. In an attempt to develop a model, two different experimental approaches are being pursued in parallel: performing single cell experiments (top down) and using biomimetic structures (bottom up), such as lipid bilayers. One challenge in many of these experiments is the reliance on fluorescent probes for detection. In the present work, we have used a label-free detection method based on an evanescent optical sensor known as an optical resonant cavity. In this approach, we are able to detect the self-assembly and solubilization of lipid bilayers in real-time. Specifically, using these silica devices, there are two independent detection mechanisms which are able to confirm the formation and detergent assisted solubilization of the lipid bilayers: 1) a refractive index change and 2) a material loss change. Both mechanisms can be monitored in parallel, on the same device, thus allowing for cross-confirmation of the results. To verify the proposed method, we have detected the formation of self-assembled phosphatidylcholine lipid bilayers from small unilamellar vesicles (SUVs) on the device surface in real-time. Subsequently, we exposed the bilayers to two different detergents (non-ionic Triton X-100 (TX-100) and anionic sodium dodecyl sulfate (SDS)) to initiate solubilization, and this process was also detected in real-time. After solubilization, the device returned to its initial state, exhibiting minimal hysteresis. The experimental wash-off was also collected and analyzed using dynamic light scattering (DLS).

Copyright © American Institute of Chemical Engineers. All rights reserved.

416775 Biosensing Mechanism of Floating Gate Transistors Utilizing Electrolyte Dielectrics

Abstract:

BIOSENSING MECHANISM OF FLOATING GATE TRANSISTORS UTILIZING ELECTROLYTE DIELECTRICS

Scott P. White, Kevin D. Dorfman, C. Daniel Frisbie,

Department of Chemical Engineering and Materials Science

University of Minnesota, Minneapolis, MN 55455, USA

In a recent publication,[1] we reported a new label-free biosensing strategy based on floating gate transistors (FGTs) that uses only solution processed, electrolyte dielectrics rather than vacuum deposited solid dielectrics. The device uses the benchmark organic polymer poly(3-hexylthiophene) or P3HT, in conjunction with an ion-gel that permits low-voltage operation. Our previous experiments demonstrated high sensitivity for DNA hybridization, which we hypothesized was due to changes in the surface potential upon binding of the analyte. In this presentation, we show that our device is sensitive to changes in both the interfacial capacitance and the surface potential of the active sensing area. This new insight allows the electrolyte dielectric FGT strategy to be readily applied to non-charged analytes (in contrast to previous FGT biosensors) while also simplifying the design of capture molecules.

To demonstrate the ability to sense interfacial capacitance, we designed a set of experiments where we systematically changed the interfacial area between the gold electrode (the sensing area) and the ion-gel. The total interfacial capacitance can thus be tuned and correlated to: (1) the current through the semiconductor and (2) the voltage when the semiconductor spikes in conductivity, V_T. When the interfacial area adjacent to the semiconductor is altered, we show that the device response can be predicted by a simple model of two capacitors in series. However, when the interfacial area of the sensing surface (decoupled from the semiconductor) is altered it results in large changes in output due to charge storage in the floating gate. These experiments and the corresponding scaling laws that we have developed highlight differences between operating with electrolyte over solid dielectrics and outline the capacitive component of the sensing mechanism.

To demonstrate the different sensing responses for capacitance and surface potential, we adjusted the work function of the floating gate electrode by selectively immobilizing self-assembled monolayers (SAMs) onto a single electrode/ion-gel interface through microfluidic channels. The SAMs were formed through thiol/gold chemistry but with a varied functional group chemistry that changes the work function from -250 to +320 meV when measured in an inert environment. The effective V_T of the semiconductor changes significantly when SAMs are formed with (1) a direction highly dependent on the orientation of the SAM but (2) largely independent on the functional group chemistry. These results are explained using a simple thermodynamic model with the high ionic strength of the ion-gel (~3.8 M) screening the electric field of the SAM functional group dipole.

1. White, S.P.; Dorfman, K.D.; Frisbie, C.D. Label-Free DNA Sensing Platform with Low-Voltage Electrolyte-Gated Transistors. *Anal. Chem.* **2015,** *87* (3), 1861-1866

Copyright © American Institute of Chemical Engineers. All rights reserved.

418082 Electrochemical Piezoelectric Cantilever Sensors: Vibration-Induced Release from Biosensor Surface

Abstract:

Objectives. The objective of this study is to develop a sensor that uses the principles of cantilever mechanics and electrochemical impedance spectroscopy (EIS) for detecting a biological analyte, and to subsequently release it through cantilever resonance vibration. The 'de-binding' technique can be used as a confirmation step, or to regenerate the sensing surface. The model analytes used in this study to demonstrate the principles are: Protein G and *E. coli* O157:H7.

Methods. Piezoelectrically-excited cantilevers were designed with a sputtered gold electrode to facilitate impedance spectroscopy measurements at the cantilever surface. Impedance spectroscopy was used in tandem with classic dynamic cantilever resonance for detecting target analyte; the measured resistance to charge transfer from the electrode to ferricyanide in solution, and cantilever resonant frequency served as the measured signals of the two techniques, respectively. All resonant frequency and EIS measurements were conducted within a specially-designed flow cell, which incorporated a reference electrode, platinum counter electrode, and the sensor gold surface as a working electrode. After Protein G or *E. coli* O157:H7 were attached to the sensor surface, the applied excitation potential which induces vibration due to the bonded piezoelectric material to the glass cantilever, was increased until the surface acceleration was sufficient to release the target analytes. Finite element modeling was employed to evaluate the relative surface accelerations for various dynamic modes present in liquid.

Results. The immobilized Protein G or *E. coli* cells were removed from the sensor electrode when the excitation potential was increased from 100 mV to 10 V, with the assistance of a negative DC potential applied to the electrode. The resistance to charge transfer decreased by ~ 40% when the 10V excitation potential was applied to a sensor saturated with Protein G with the assistance of a -700 mV DC potential applied to the electrode. Little to no removal of analyte was observed with the DC potential alone. The signal returned to its original value when the surface was exposed to fresh Protein G, indicating regeneration of the sensing surface.

Conclusion. . The vibration of the piezoelectric sensor triggered release of the physisorbed Protein G and antibody-bound *E. coli* with the assistance of electrostatic repulsion. The vibration-induced release approach is useful for repeated sensing requirement applications, and can be used to separate dilute analytes from complex matrix. We show in this study that sensor surface can be regenerated.

Copyright © American Institute of Chemical Engineers. All rights reserved.

(275a) Nanosensors-Encapsulated Microcapsule for Glucose Sensing

Xie, X., Massachusetts Institute of Technology
Zhang, W., Harvard University
Abbaspourrad, A., Harvard University
Weitz, D. A., Harvard University
Anderson, D., Hope College
Originally presented on: 11/10/2015 8:30:00 - 8:52:00

Non-invasive implantable systems that are able to continuously track blood glucose levels are needed for diabetes diagnostics and monitoring. Although many nanomaterial colloidal solutions have been developed as optical sensors to detect glucose in aqueous samples, the liquid nature of these sensor solution limits their applications as implantable devices for glucose sensing in vivo. One potential solution is embedding the nanosensors in biocompatible hydrogel matrix, yet the direct contact between the sensor surface and the polymer chains of the hydrogel tends to disrupt the performance and sensitivity of the sensors. In this work, glucose responsive nanosenors were encapsulated in liquid-core microcapsules. Glucose is allowed to diffuse through the capsule shell freely, while the nanosensors are trapped inside the capsules without leaking away. The interaction between glucose and the nanosensors produces detectable optical signals. The microcapsules were fabricated by forming water-oil-water template double emulsion using microfluidic techniques. The capsules' stability and permeability to glucose were demonstrated. Three types of glucose-responsive nanosensors, glucose oxidase (GOx)-functionalized quantum dots, GOx-functionalized gold nanorods, and boronic acid-functionalized carbon nanotubes, were encapsulated in the microcapsules individually, and they responsively exhibited optical signals in the present of glucose. Furthermore, implantable devices were fabricated by embedding these capsules in calcium alginate hydrogel, and their responses to glucose were demonstrated.

430221 Protein Efflux Mapping in Single Living Cells with Synthetic Optical Nanosensors

Abstract:

Recent advances in metabolic engineering of *Escherichia coli* have provided an abundance of platforms for natural product biosynthesis and protein export [1]. However, real-time optical detection of *E. coli* metabolites remains a challenge. Previous work has shown great promise for the use of near-infrared fluorescent nanomaterials for optical detection of biomolecules [2, 3]. We present a platform for near-infrared optical detection of proteins exported from *E. coli*. By electrostatic conjugation of an aptamer-anchor polynucleotide sequence to a single-wall carbon nanotube, we are able to detect both RAP1 GTPase and HIV-1 integrase proteins, which each produce increases in near-infrared fluorescence of their respective aptamer-SWNT conjugates. Herein, we engineer the RAP1 gene, which is native to yeast, into *E. coli*. Subsequently, we monitor the productivity of RAP1 protein export by individual engineered *E. coli* cells over the course of 1 hour. We show that the process of induction, protein synthesis, and protein export is a highly stochastic process yielding variability in protein secretion on a single-cell level. We further show that our platform can be extended to the detection of protein from *E. coli* infected with T7 bacteriophages carrying the RAP1 gene. Our platform shows promise of extension for single-cell optical detection of a broad range of metabolic products with high spatial and temporal resolution.

1. Keasling, J.D., *Synthetic biology and the development of tools for metabolic engineering.* Metabolic Engineering, 2012. **14**(3): p. 189-195.
2. Landry, M.P.*, Kruss, S.*, et al., *Neurotransmitter Detection Using Corona Phase Molecular Recognition on Fluorescent Single-Walled Carbon Nanotube Sensors.* Journal of the American Chemical Society, 2014. **136**(2): p. 713-724.
3. Landry, M.P.*, Zhang, J.Q.*, et al., *Molecular recognition using corona phase complexes made of synthetic polymers adsorbed on carbon nanotubes.* Nature Nanotechnology, 2013. **8**(12): p. 959-968.

Copyright © American Institute of Chemical Engineers. All rights reserved.

433673 Advances in SERS Based Point of Care Diagnostic Assays

Abstract:

In the development of next generation bioassays designed for detecting the emergence and spread of diseases in point-of-care (POC) and remote settings, more rapid and portable analytical methods are required. Colloidal based SERS bioassays have distinct advantages for use in portable diagnostic systems over surface sensors. Nanoparticles in solution allow for both a freely diffusing sensor surface and a high surface area to volume ratio. Therefore colloidal systems are not impacted by mass transfer limitations, allowing for the rapid analysis of small sample volumes. In addition, paramagnetic capture assay schemes provide both separation and concentration mechanisms, minimizing signals from non-specific interactions. Recent developments in our laboratory in the fabrication of gold-coated magnetic nanoparticles provide an additional alternative single nanoparticle assay scheme. The nanoparticles function as SERS substrates while simultaneously allowing for manipulation by a magnetic field. Detection strategies have been demonstrated for the sensitive detection of Rift Valley Fever virus and West Nile virus nucleic acids and proteins in multiplexed formats. In addition, we have demonstrated the advantages of using antibody fragments over IgGs as a more compact recognition element in our immunoassay schemes.

Copyright © American Institute of Chemical Engineers. All rights reserved.

434940 Functionalization of Nanostructured ZnO Surfaces with Selective Linker Molecules Towards Biomolecule Detection

Abstract:

In this study, the influence of surface functionalization on charge transfer in nanostructured zinc oxide (ZnO) surface was investigated. Sensing sites with tailored surfaces offer novel immobilization strategies for selective functionalization and specific interaction of protein biomolecules, which can lead to enhancement in specific output signal response and hence selectivity. Functionalization of nanostructured ZnO surfaces were evaluated using two organic molecules with thiol and phosphonic acid functional groups. Dithiobis(succinimidyl propionate) (DSP) and 11-Aminoundecylphosphonic acid (AUPA) covalently link to ZnO providing NHS binding sites for controlled immobilization of antibody. Fluorescence intensity measurements with varying concentrations of DSP and AUPA labeled with Rhodamine B were performed, to identify the amount of functionalization. We observed saturation in fluorescence intensity with increasing concentration of linker molecules. Such modified ZnO surfaces were characterized using contact angle measurements, X-ray photoelectron spectroscopy (XPS) and electrically using electrochemical impedance spectroscopy (EIS) and Mott-Schottky (MS) analysis. XPS spectra revealed selective binding of thiol to zinc interstitials and phosphonic to oxygen vacancies on nanostructured ZnO. To demonstrate the potential of functionalized surfaces for biomolecule detection, cardiac troponin-T, a protein biomarker specific for acute myocardial infarction was chosen. The influence of selective functionalization on charge transfer and perturbations at ZnO electrode-electrolyte interfaces due to biomolecular binding was studied using EIS and MS analysis. Our results demonstrate that leveraging zinc interstitials increases the sensitivity of biomolecule detection over larger dynamic range and detection limit at 10 fg/mL.

Copyright © American Institute of Chemical Engineers. All rights reserved.

433015 Using Phage and Phage Genomes for Generating Highly Amplified Signals for Protein Sensing

Abstract:

For early detection of many diseases, it is critical to be able to diagnose small amounts of biomarkers in blood or serum. One of the most widely used sensing assays is the enzyme-linked immunosorbent assay (ELISA), which typically uses detection monoclonal antibodies conjugated to enzymes to produce colorimetric signals. In order to increase the overall sensitivities of these sensors, in the first half of the talk, I will describe our use of a dually-modified version of filamentous bacteriophage that produces significantly higher colorimetric signals in ELISAs than what can be achieved using antibodies alone. Since only a few proteins at the tip of the micron-long bacteriophage are involved in antigen binding, the ~4000 other coat proteins can be augmented - either by chemical functionalization or genetic engineering - with hundreds to thousands of functional groups. In this work, we demonstrate the use of bacteriophage that bear a large genomic fusion which allows it to bind specific antibodies on coat protein 3 (p3) and multiple biotin groups on coat protein 8 (p8) to bind to avidin-conjugated enzymes. In direct ELISAs, the anti-rTNFa conjugated bacteriophage show about 3 to 4-fold gains in signal over that of anti-rTNFa, demonstrating their use as platforms for highly sensitive protein detection.

In the second half of the talk, I will highlight our recent efforts of using the genetic information of the M13 bacteriophage to generate highly amplifiable signals in a single solution isothermally. While one method for producing amplified signals in the presence of antigens has been to use oligonucleotide-conjugated antibodies followed by real-time or isothermal PCR to amplify the DNA strands, the DNA conjugation methods used can affect an antibody's ability to bind its protein target, and many rounds of linear amplification are often required for sufficient signal generation. In contrast, because the viruses naturally contain a large, 9.5 kilobase genome, combining isothermal rolling circle amplification (RCA) with multiple priming sites can yield statistically significant positive signals at sub-femtomolar antigen concentrations. The phage therefore provides a convenient container to carry the detection genome to the analyte. I will show that since the genome can be released without or minimal heat denaturation of the phage, and that the number of DNA amplification steps (primer annealing, extension, and dehybridization) are limited, massive amounts of DNA can be detected in real time by luminescence in a much shorter time, in a single reaction tube, at room temperature, and without a specialized PCR instrument. I will also show that the virus genome can be easily modified with restriction sites at different locations so different analytes can be identified through digestion with a single restriction enzyme and gel electrophoresis.

Copyright © American Institute of Chemical Engineers. All rights reserved.

413030 Multiplexed Biomarker Detection Using Phase Change Nanoparticles

Abstract:

Many biomarkers are specific for early stage cancer detection, but no individual biomarker is ideal to distinguish lethal cancer from indolent ones due to lack of tumor specificity. Parallel to efforts of finding specific biomarkers, a feasible way of providing better predictive value is to detect multiple biomarkers from a sample and collectively assess response pattern of more biomarkers. But, existing techniques such as microarrays and nanoparticles are not suitable to detect multiple low concentration biomarkers in a tiny amount of sample due to low sensitivity, low multiplicity and extensive sample preparation. We have detected multiple DNA and protein biomarkers with a panel of phase change nanoparticles of metals and alloys. Nanoparticles attached on a substrate through ligand-biomarker interactions are readout using differential scanning calorimetry, where the peak position and area reflect presence and concentration of biomarkers. This research is innovative for several reasons. (1) It uses an unexplored phenomenon of solid materials, i.e., the temperature of a solid will not rise above its melting point until the entire solid is molten. (2) Sharp melting peak, large thermal scan range and wide choices of materials enhance multiplicity this detection without adding system complexity. (3) The surface grafting density of ligands on nanoparticles is controlled to detect multiple biomarkers with concentrations differing several orders of magnitude.

We have identified ten different metals that form binary eutectic alloys among any two of them are identified from phase diagram database. These metals (i.e., aluminum, bismuth, cadmium, copper, gadillium, indium, lead, magnesium, palladium, silver) can form 45 binary eutectic alloys, 120 ternary eutectic alloys, 210 quaternary eutectic alloys, and so on. The total number of metals and eutectic alloys are 1,023. Pandat 8 software (CompuTherm) has been used to derive compositions of ternary and higher order eutectic alloys formed by these ten metals. Two thermophysical properties (melting temperature and latent heat of fusion at eutectic composition) had been extracted from data. After obtaining eutectic compositions, nanoparticles had been made by thermally decomposing of organometallic precursor in a high boiling point solvent (ethylene glycol) by using polyvinyl alcohol (PVA) as surfactant. We have used these nanoparticles to detect proteins and single strand DNA (ssDNA).

Copyright © American Institute of Chemical Engineers. All rights reserved.

424157 Selective Vaporization of Superheated Nanodroplets for Rapid, Sensitive Acoustic Biosensing

Abstract:

This work describes a simple, scalable ultrasound imaging detection method in which aggregation of superheated droplets allows detection of small concentrations of biomolecules without the need to wash out excess reagent. While many contrast agents can be functionalized with ligands that can bind membrane-bound targets, fewer agents are able to sense small quantities of soluble, nonreactive biomarkers and relay their location using a modality capable of deep-tissue imaging. The mechanism of biosensing in this work derives from the discovery that biomolecule-specific aggregation of superheated nanodroplets greatly changes their ability to be vaporized into microbubbles. While in their liquid form the nanodroplets possess poor acoustic contrast properties owing to their incompressibility, these superheated nanodroplets can be vaporized using High Intensity Focused Ultrasound (HIFU), resulting in the generation of high-contrast microbubbles in situ. We show that even small-order aggregates of droplets joined via biotin-streptavidin interactions can specifically be vaporized using HIFU, thus allowing sensing of analytes with a dynamic range of 100 fM to 100 nM. Because of the bulk droplet aggregation mechanism of detection, full activation of signal can be achieved within 10 min. Finally, this work provides evidence of a mechanism for aggregation-based droplet vaporization that will aid the design of future ultrasound contrast agents and biosensors for in vivo imaging.

Copyright © American Institute of Chemical Engineers. All rights reserved.

419285 A Wearable Integrated Sensor Array for Perspiration Analysis

Abstract:

Wearable technologies can play a significant role in realizing personalized medicine through continuously monitoring an individual's physical and physiological states. Most currently developed wearable technologies are capable only in tracking the physical activities of an individual and fail to provide insight into the individual's state of health. Human sweat contains the physiologically rich information needed to infer an individual's sate of health and is an excellent candidate for non-invasive monitoring. The wearable sweat sensors can serve as an ideal platform for a wide range of real-time healthcare monitoring such as exercise-induced dehydration and medical diagnosis.

Copyright © American Institute of Chemical Engineers. All rights reserved.

428893 Curved and Self-Folded Polymer-Graphene Sensors

Abstract:

Graphene has exceptional electronic, mechanical and optical properties of relevance to sensing applications. In order to create three dimensional curved and flexible sensors we describe the use of strain engineering in graphene/SU-8 bilayers to form curved and folded sensors and discuss applications as stimuli responsive materials, electronic and optical sensing.

Copyright © American Institute of Chemical Engineers. All rights reserved.

433911 Paper-Based Home Biosensor for Detecting Phenylalanine from a Sample of Whole Blood

Abstract:

Biosensors are important tools for measuring concentrations of specific analytes found in biological fluids. Paper-based microfluidic devices can offer many benefits to the field of biosensor development. Paper-based materials are inexpensive, easily disposable, and readily available. Small volumes of fluid can be used in these devices, which decrease the need for costly reagents. And capillary flow in paper materials removes the need for pumping equipment. The topic of this presentation is a paper-based, biosensing device that can be used for therapeutic, home monitoring by people with the genetic disorder Phenylketonuria (PKU.)

People affected by PKU are required to maintain a strict, low Phenylalanine (Phe) diet to avoid irreversible, adverse health affects. Maintaining appropriate levels of Phe can be especially challenging for young children, adolescents and pregnant women. Much like a glucose test kit helps a person with diabetes maintain appropriate glucose levels, a device that can monitor the concentration of Phe in a finger prick of blood could aid a person with PKU maintain appropriate Phe levels and adhere to diet therapy. Current testing for Phe is laboratory-based, requiring expensive equipment or trained technicians.

Our proposed solution is a semi-quantitative, paper-based assay that utilizes a well-known, two-step chemical reaction to produce a colorimetric response that correlates Phe concentration to color intensity. The assay accepts a finger prick volume of blood (40 μL) into an asymmetric polymeric membrane that filters out cellular components and allows plasma to wick into two glass fiber pads where the first reaction takes place for six minutes. These pads have been impregnated with Phenylalanine Dehydrogenase (PheDH) and the oxidized form of Nicotinamide adenine dinucleotide (NAD^+.) Phe is catalyzed by PheDH and the reaction reduces NAD^+ into NADH. A third glass fiber pad, impregnated with Nitro blue tetrazolium chloride (NBT) and methoxy Phenazine Methyl Sulfate (mPMS,) is then folded to make contact with a pad containing the products of the first reaction to allow for a second reaction to take place for an additional minute. NBT is reduced by NADH from the first reaction (in a process aided by mPMS) to its formazan dye product. This visible color change is recorded with a desktop scanner. Statistically relevant intensity differences have been measured in concentrations of blood spiked with 0 (normal), 3.75 (slightly elevated) and >7.5 (substantially elevated) mg/dL Phe using this device.

The focus of this presentation will be on optimization of the flow path of reagents to achieve maximum signal-to-noise ratio in the detection region by controlling the patterning and drying of reagents in the multi-component assay.

Copyright © American Institute of Chemical Engineers. All rights reserved.

429106 Novel Electrical Impedance System for Rapid Detection of Mycobacteria

Abstract:

Tuberculosis (TB) is a common infectious disease of the human respiratory system caused by *Mycobacterium tuberculosis*. While it has a high mortality rate, it can be treated with timely detection. Hence the rapid detection of *mycobacteria* is an essential component of efforts to control the spread of TB. The current gold standard for detection of TB (especially in low-resource environments) involves culture-based diagnostic techniques. These techniques are able to infer the presence of *live* microorganisms in a sample by detecting changes to the properties of the suspension (such as O2/CO2 levels, pH, electrical conductivity, temperature etc) brought about by bacterial metabolism. Although these methods are relatively inexpensive and reliable, they are time consuming. For slow growing *mycobacterium*species, which have a doubling time of ~24 hours, this method of detection can take anywhere from 7-56 days. The objective of our research is to reduce the time to detect mycobacterium in a sample and thereby prevent the delay in treatment.

We achieve this using a multi frequency electrical impedance spectroscopy (EIS) to detect the presence of mycobacterium in a sample that has been treated to eliminate all organisms except mycobacteria. Under the influence of an AC field, the membrane of a *living* cell, which encloses a cell at a different potential than that of its surroundings (membrane potential), acts as a capacitor. Thus the "bulk capacitance" of a suspension is proportional to the number of living cells present. Consequently, any increase in the bacterial number is reflected as an increase in the bulk capacitance of the system – which we monitor over time. A "significant" change in the value of the bulk capacitance from its baseline (time t=0) value indicates the presence of growing microorganisms (presumably mycobacteria) in the sample of interest.

This basic approach (multi-frequency EIS) has been previously used by our group to perform blood cultures (wherein blood suspected of being infected with bacteria is added to growth media and the presence of live organisms is inferred) significantly faster than currently methods that rely on bacterial metabolism. In this study, we demonstrate the applicability of this technique towards detecting *mycobacteria*, and obtain estimates of the amount of time that could be saved in detecting*mycobacteria* upon using our method.

For our study, we used *Mycobacterium bovis* (a zootonic analog of *M. tuberculosis*). *M. bovis* was cultured in modified Middlebrook 7H9 media, and samples were prepared with three different initial loads of M. bovis: "low" (2-5 CFU/ml), "medium" (100-200 CFU/ml) and "high" (~3000 CFU/ml). Each sample (and a control) was assayed using both the established technique (BACTEC MGIT ™ culture) and our method, which involves periodic (once every 12-24 hrs) sampling of the culture fluid, measurement of electrical impedance at frequencies ranging from 10KHz to 100 MHz, an analysis of EIS data to calculate the bulk capacitance of the sample, and tracking the calculated bulk capacitances over time.

We found that for control samples, the bulk capacitance of the suspensions did not change in a statistically significant manner from their baseline values. On the other hand, the bulk capacitance values different samples containing *Mycobacteria* increased over time and become statistically significantly different from baseline when concentrations rose to ~ 10,000 CFU/ml. Thus our times to detection (TTDs) were about 60-70 hrs (< 3 days) for "high" concentration samples, 84-108 hrs (< 5 days) for "medium" concentration samples, and about 132-156 hrs (< 7 days) for "low" concentration samples. In contrast the TTDs for the laboratory standard was ~ 6 days for "high" concentration samples, and > 40 days for "medium" and "low" concentration samples (we are still awaiting results from the University Hospital microbiology lab).

Hence our method promises to deliver results significantly faster than current automated systems used in commercial clinical microbiology laboratories.

Copyright © American Institute of Chemical Engineers. All rights reserved.

ARIZONA STATE UNIVERSITY

The Development of a New Differential Pressure Flow Meter for Bidirectional Measurement of Human Breath Flow

Devon Bridgeman

ASU

ARIZONA STATE UNIVERSITY

Applications of Flow Measurement in Human Breath

- Two main types of breath analysis
  - Forced
    - Spirometry, patient is asked to exhale as hard as possible
  - Relaxed
    - Most methods, patient breaths at a natural or close to natural pace

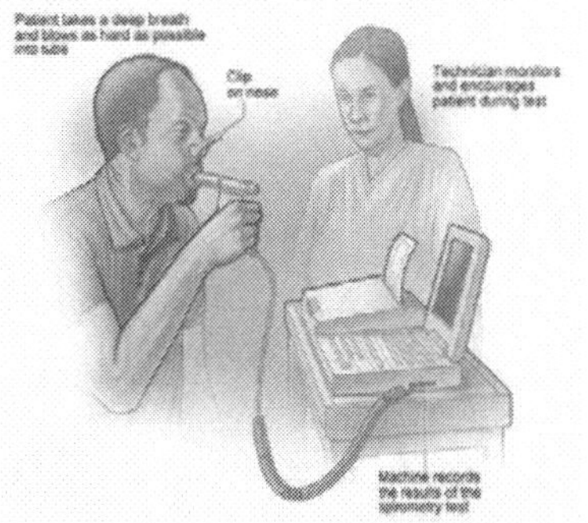

1

ASU

The Issue

- With any flow measurement, there is a tradeoff between the accuracy, resistance, and cost
  - For any clinical or consumer application, cost is a large factor

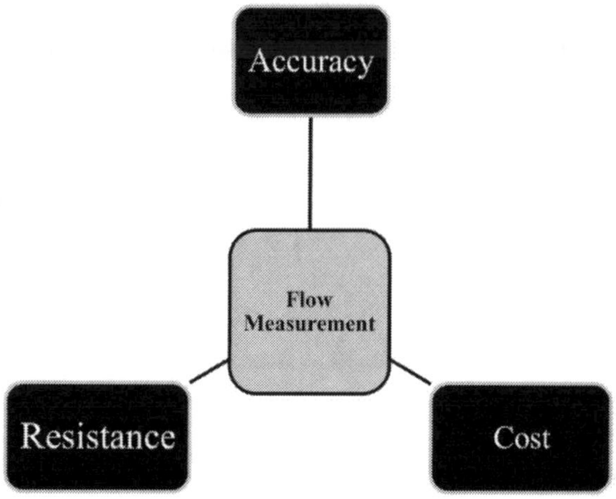

2

ARIZONA STATE UNIVERSITY

Human Breathing

- Added challenges with breath
  - Very dynamic measurement
  - High sensitivity to flow-resistance
- Even more added challenges with rebreathing
  - Periodic temperature swings
  - Periodic humidity swings
  - Directionality
  - Higher sensitivity to flow-resistance

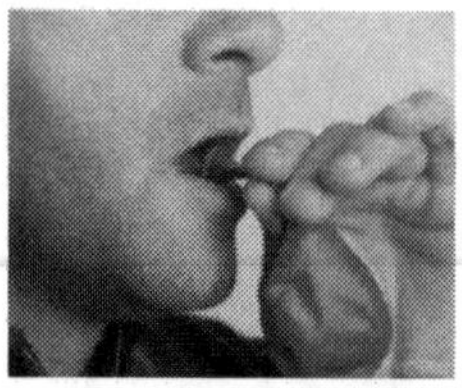

3

ARIZONA STATE UNIVERSITY

Differential Pressure Flow Designs

- For low-cost monitoring, differential pressure is attractive
  - Often fixed- or variable-orifice meters are used for relaxed flows
    - Rough tradeoff between sensitivity and resistance with fixed-orifice designs
    - Wear issues with variable-orifice designs
  - Fleisch meters often used in clinical settings
- Pitot Tubes not seen in this type of application

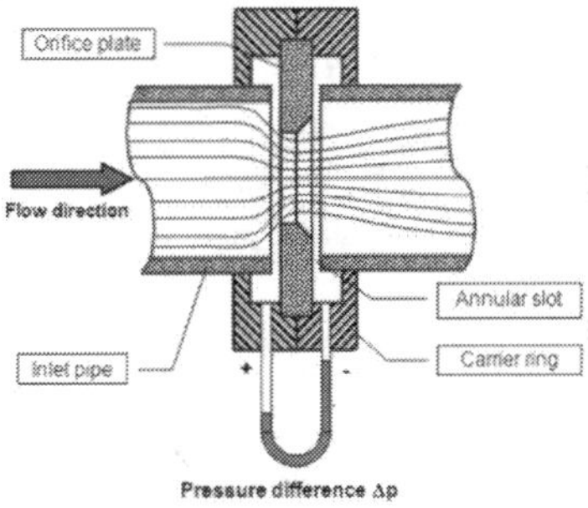

Sagemetering.com

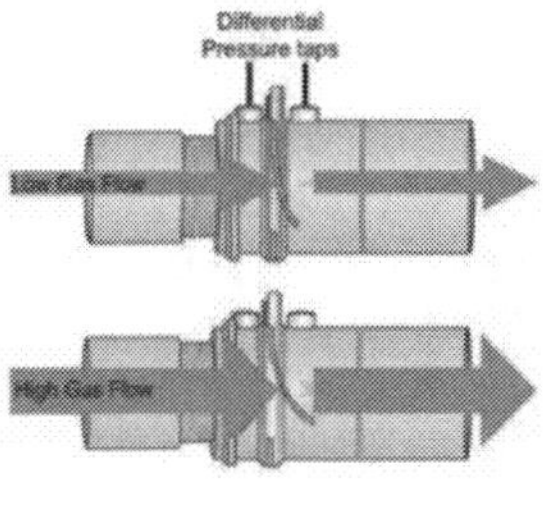

apsf.com

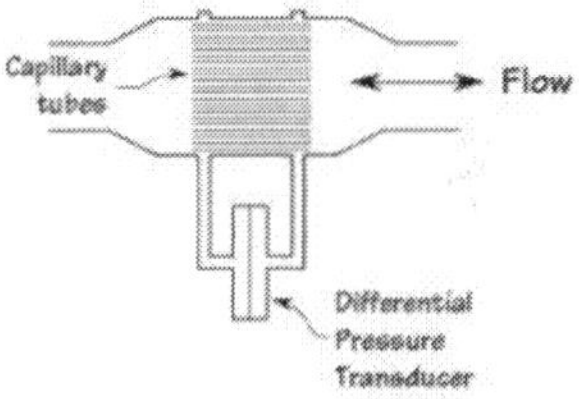

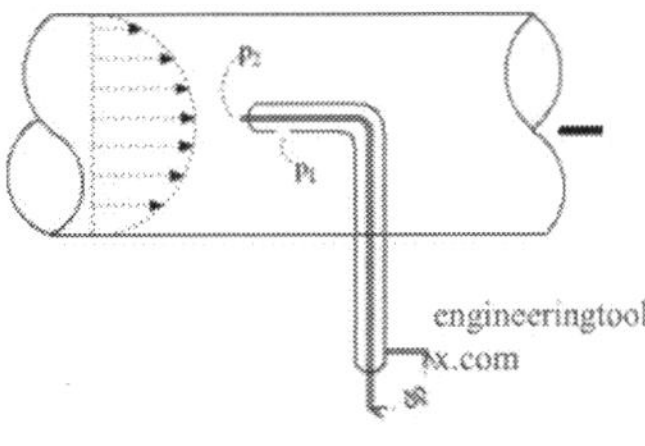

engineeringtoolbox.com

4

ASU

ARIZONA STATE UNIVERSITY

The Confined Pitot Tube

- The Confined Pitot Tube (CPT) flow meter combines multiple effects
  - Stagnant Pressure
  - Confinement
  - Inertial Effects

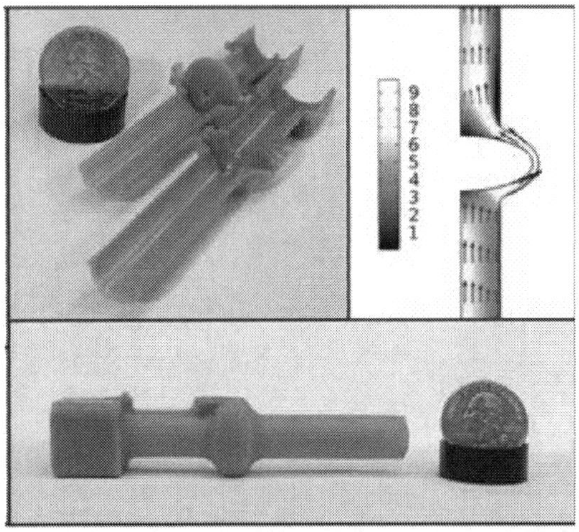

5

ASU

Pressure Distribution

- Pressure profile shows a large stagnant pressure in the center of the disk
- Negative pressure after the bend
- Two options for sampling
  - 1-Point
  - 2-Point

P1 Vs. P2

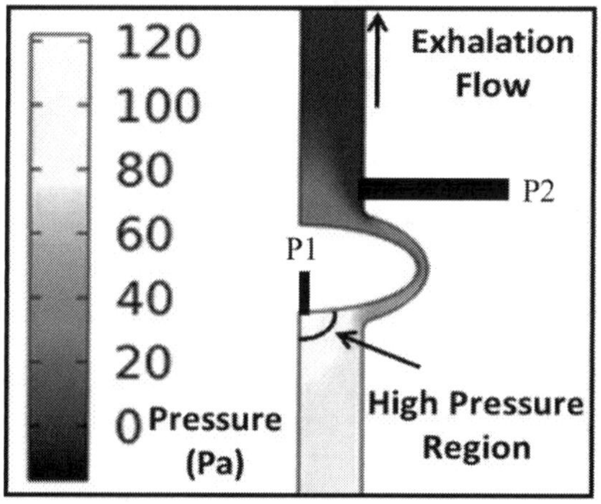

Flow Meter Comparison

Fixed Orifice-Plate

Pitot Tube

CPT (High-Confinement)

CPT (Low-Confinement)

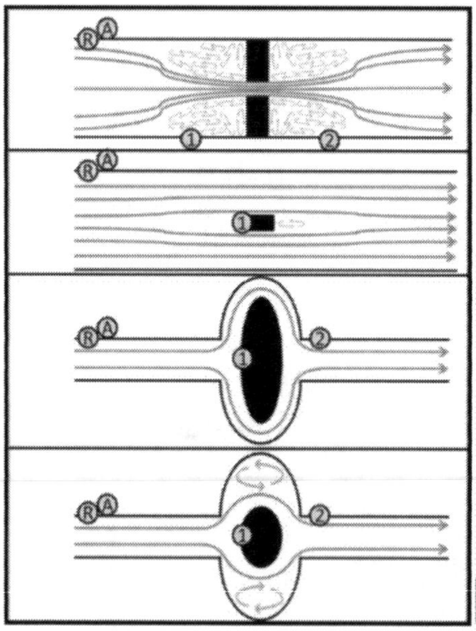

Comsol® CFD Simulations

- Simulations were run in Comsol Multiphysics to determine an optimal design
  - Non-trivial transitions between laminar and turbulent flow
  - Solution: Run the simulations under both sets of physics and compare experimentally
- A range of inner disc sizes probed

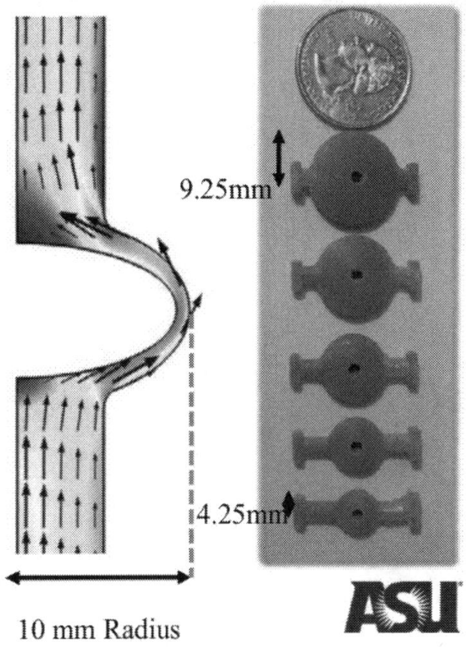

ASU

ARIZONA STATE UNIVERSITY

Static Flow Tests

- Tests done at constant flows of dry compressed gas
- Sensiron flow meter used as reference

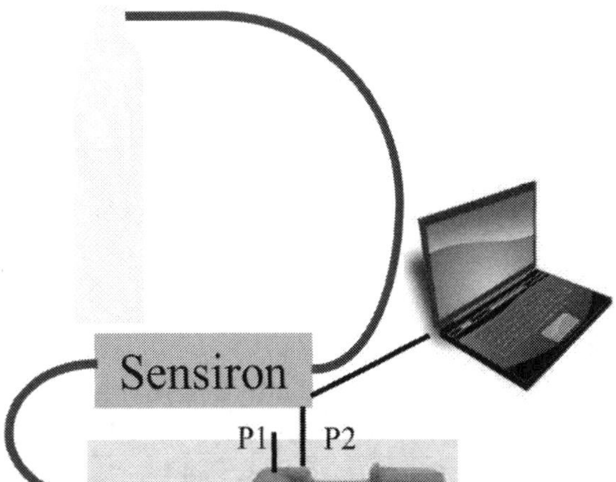

ASU

ARIZONA STATE UNIVERSITY

Experiments Vs. Simulations: Small Radii

1-Point Sampling | 2-Point Sampling

4.25 mm Disk

5.5 mm Disk

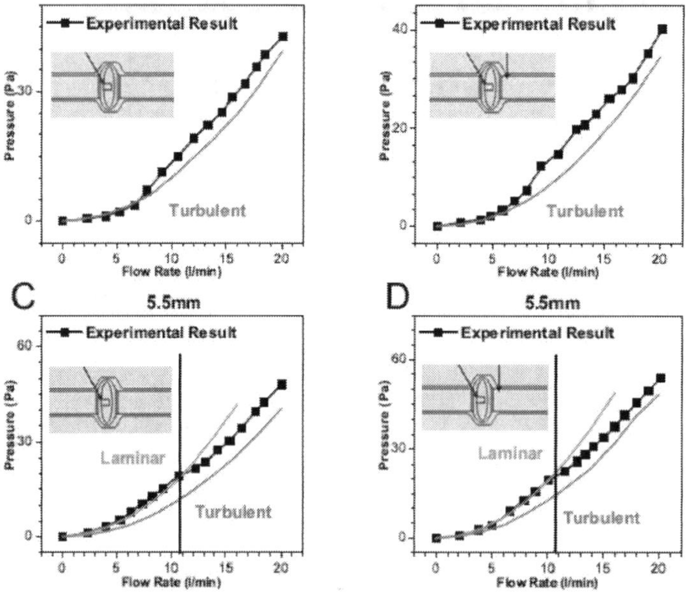

10

ARIZONA STATE UNIVERSITY

Experiments Vs. Simulations: Large Radii

1-Point Sampling | 2-Point Sampling

0.8cm Disk

0.925 cm Disk

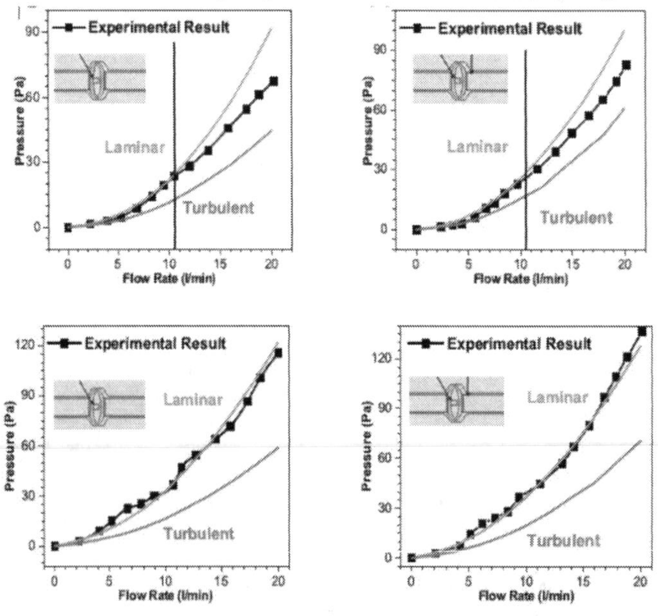

11

ARIZONA STATE UNIVERSITY

Comparison to a Fixed-Orifice

- If we compare to a fixed-orifice flow meter (2.5mm radius), we see:
  - More rapid scaling
  - Comparable performance at low flows

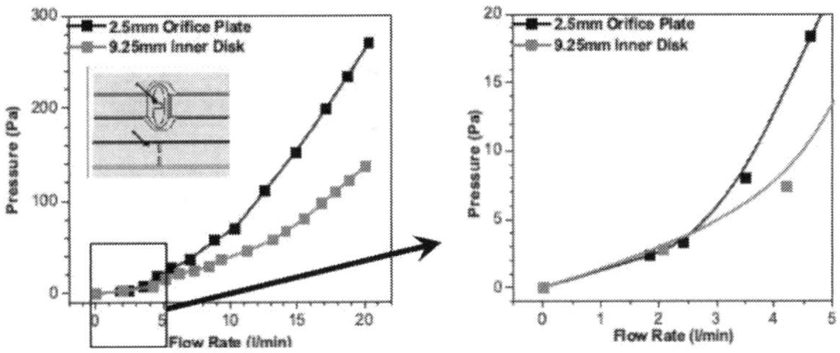

12

ARIZONA STATE UNIVERSITY

…and on the Other Hand

- The flow-resistance scales much slower with the most-confined setup
- As we go to smaller inner disk diameters, the flow-resistance drops off rapidly

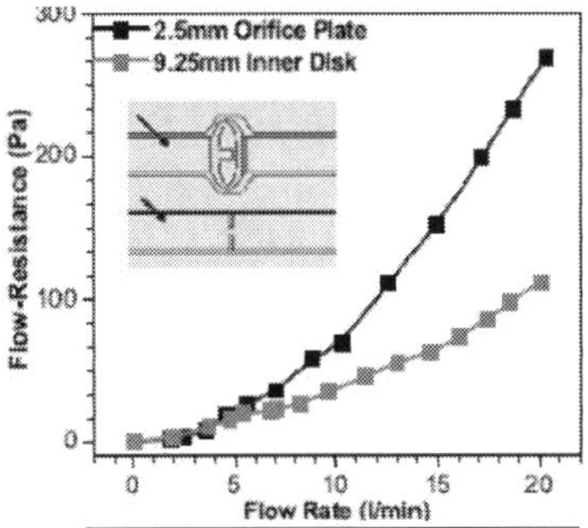

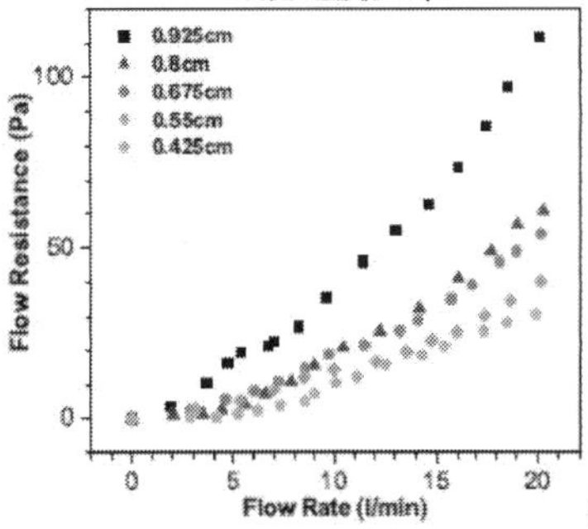

13

ARIZONA STATE UNIVERSITY

Artificial Breathing Patterns

- We wanted to probe breathing waveforms on a known platform
- Modified Vacumed® breath simulator

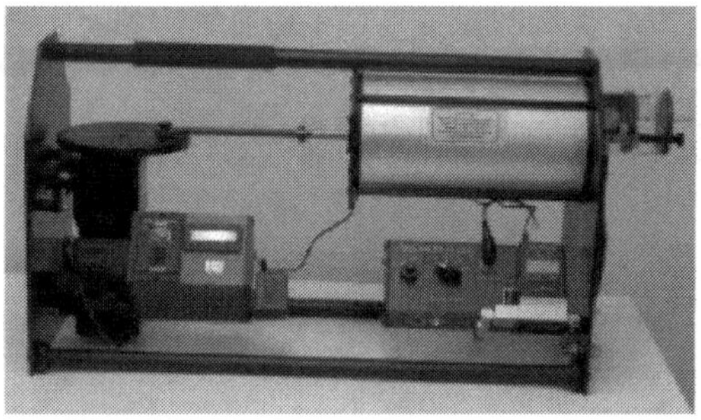

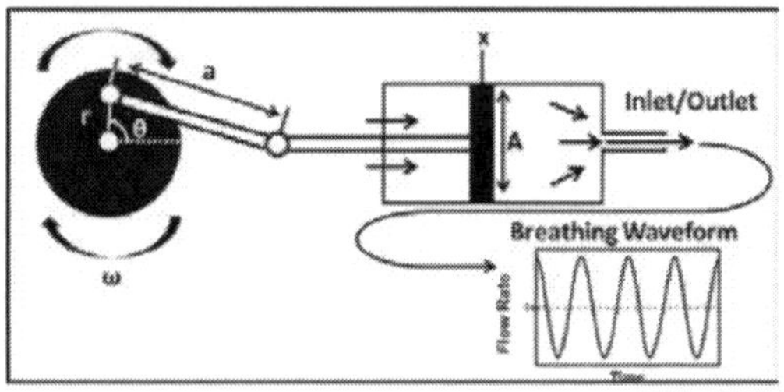

14

ASU

ARIZONA STATE UNIVERSITY

Artificial Breathing Results

- Using a multiple-equation method, we are able to read something very similar to what we expect

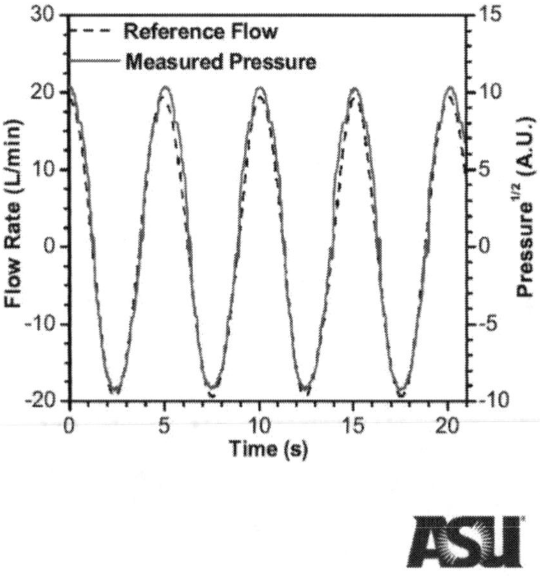

15

ASU

ARIZONA STATE UNIVERSITY

Actual Breath Tests

- Comparison to an Oxycon Mobile unit
  - Volume comparison
  - Turbine Flow meter
- Strong trend, although off by a linear factor
  - Easily Corrected!

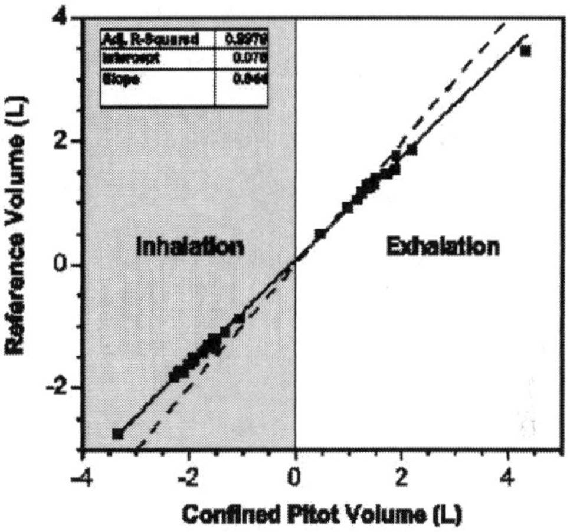

16

ASU

ARIZONA STATE UNIVERSITY

Actual Breath Test Errors

- Volume error comparison
- Standard deviation of less than 50mL!
  - Not bad when considering the published volume accuracy of the commercial unit!

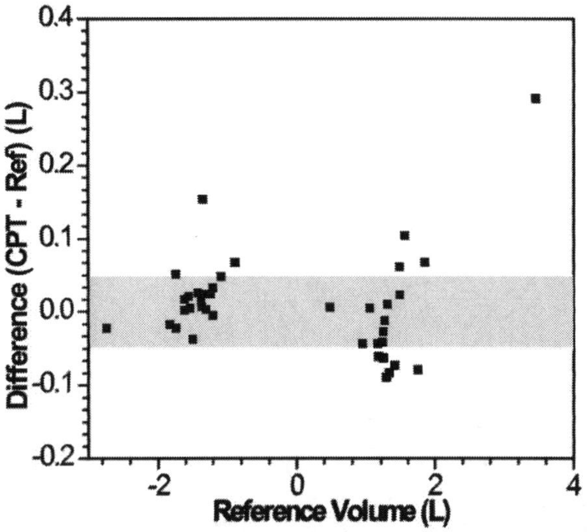

17

ASU

ARIZONA STATE UNIVERSITY

Conclusions

- The Confined Pitot Tube combines the effects of stagnant pressure, flow confinement, and inertia
- At high confinement, the flow meter displays mostly laminar behavior
- The Confined Pitot Tube provides good performance at a low flow-resistance
- The Confined Pitot Tube is suitable for measuring human breathing patterns

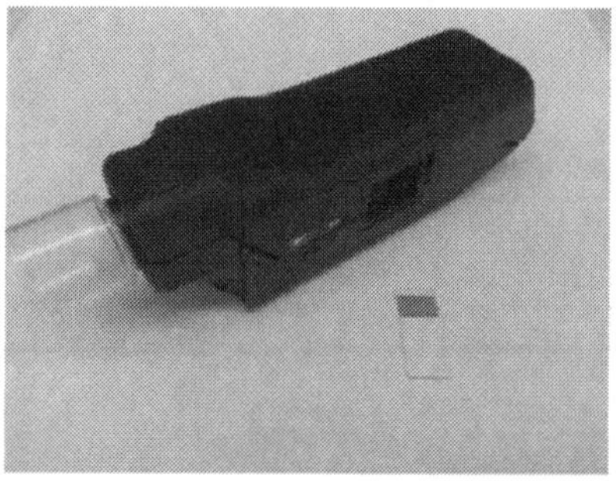

18

ASU

ARIZONA STATE UNIVERSITY

Thank You!

- I would like to thank:
  - Dr. Erica Forzani
  - Dr. Xiaojun Xian
  - Dr. Francis Tsow
  - Dr. N.J. Tao

GPSA graduate and professional student association gpsa.asu.edu

NIH

ASU BIODESIGN INSTITUTE ARIZONA STATE UNIVERSITY

19

ASU

ARIZONA STATE UNIVERSITY

Gas Properties

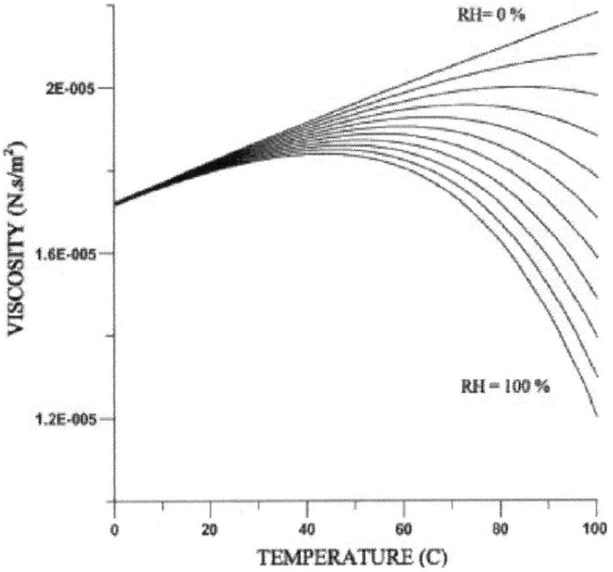

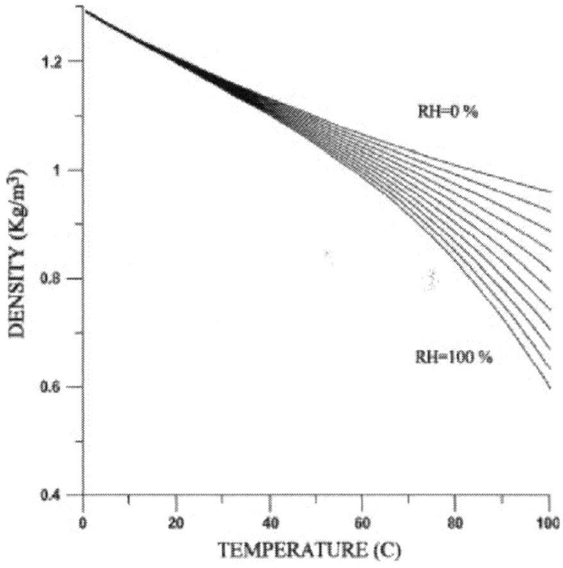

ASU

433772 Controlled ZnO Nanostructures Alignment and Electrode Spacing – an Approach Towards Multiplexed Biomarker Detection

Abstract:

We demonstrate a methodology to tailor the morphology of zinc oxide (ZnO) nanostructures growth in aqueous chemical bath which in turn dictates the multiplexing capabilities of nanostructured microelectrode sensor platform. A key challenge in multiplexed biomarker quantification is to detect low concentrations of target analytes with high specificity and sensitivity from physiological fluids. Technological advancements provide capabilities to synthesize and integrate nanomaterials into high density arrays leading to development of miniaturized, portable electrical biosensor platforms that meets these requirements. In this work, we demonstrate the multiplexed electrical detection of cardiac biomarkers, troponin-T (cTnT) and troponin-I (cTnI) using one dimensional (1D) ZnO nanostructures as our sensing platform. Electrode characteristics such as material, dimensions and its surface modifications dictate the performance of electrochemical biosensors. In this regard, we demonstrate the methodology to design ultra-small electrodes using 1D ZnO nanostructures via hydrothermal synthesis. We have studied and identified the influence of various aqueous chemical bath parameters on morphology of ZnO nanostructure formation and characterized using SEM and AFM. We have established previously that the density of ZnO nanostructure on microelectrode platform can be effectively controlled by varying precursor concentration, which in turn modulates the electrode behavior (micro/ nanoelectrode). Nanoelectrode behavior enables biomolecule detection under steady state condition due to improved electrochemical reaction rates and increased signal to noise ratio. Here, we discuss the various electrode configurations that enables detection of multiple analytes simultaneously in physiological buffer. Electrode spacing and dimensions that offer minimum interconnect issues and no signal crosstalk were identified through COMSOL simulation and verified experimentally. The charge perturbations due to biomolecular binding events on nanostructured sensing sites were recorded using electrochemical impedance spectroscopy (EIS). Results demonstrate amplified signal response and larger dynamic range for cTnT and cTnI detection with limit at femtogram/mL. Multiplexed detection of these two biomarkers yield about 90% sensitivity and selectivity that individual detection of cTnT and cTnI on separate microelectrode platform. Concurrent detection of these cardiac biomarkers have greater significance in improving quality of life for those suffering from acute myocardial infarction.

Copyright © American Institute of Chemical Engineers. All rights reserved.

425292 Developing Low-Cost Nanowire Sensors Based on a Seed-Mediated Solution Process

Abstract:

This talk describes the translation of research discovery and patented technology made in our lab into nanosensor devices. Nanowires have been applied to sensing for over 10 years but few nanowire sensors have reached the market. The major barriers are the complexity of manufacturing and difficulty in connecting nanowires in devices. Our technology can potentially overcome these barriers by using a different approach and with materials not previously used as nanowires. We discovered that when mixing organic crystalline compounds with nanoparticles in a solution, nanowires nucleated on nanoparticles upon solvent evaporation. We developed and tested a working hypothesis that links the nanowire formation to the nanoconfinement effect imposed by the nanoparticle seed. The nanoconfinement concept has been successfully applied to the synthesis of nanowires of conductive tetrathiafulvalene charge-transfer and potassium tetracyanoplatinate salts on gold nanoparticle seeds using electrocrystallization. We further synthesized the organic nanowires directly on microelectronic substrates by using features of the substrate pattern as nucleation points for nanowire formation. We tested the nanowire device for electrochemical sensing of different vapors. This research contributes new nanowire materials as signal transducers in nanosensor devices built on commercial electronic components.

Copyright © American Institute of Chemical Engineers. All rights reserved.

433232 Sensor Data Analysis for Environmental Exposure Assessment

Abstract:

The exposome represents the totality of exposures from conception onwards, simultaneously identifying, characterizing and quantifying the exogenous and endogenous exposures and modifiable risk factors that predispose to and predict diseases throughout a person's life span. Unraveling the exposome implies that both environmental exposures and genetic variation are reliably measured simultaneously. The spread of smartphone applications and fitness monitors provides new and less expensive methods for tracking time-location-activity patterns in exposure studies. Small sensor devices and smartphone apps can help exposure assessors to determine both location and activity of individuals with greater ease. As part of the Health and Environment-wide Associations based on Large population Surveys (HEALS) research project, this study examines the feasibility of using a series of sensors for tracking personal location and activities. This is the first step of the development of a methodological approach to estimate the exposome, which encompasses the totality of human environmental exposures at an individual level.

25 participants in the city of Thessaloniki, Greece during a summer week, wore a series of devices such as

(a) a temperature logger to detect changes between indoor and outdoor conditions, (b) a commercially available fitness monitor to capture motion and intense of activity, (c) a GPS device to track location and speed along with (d) Moves, a smartphone application that enables tracking of location and activity. Additionally, a time activity diary was filled out on paper by participants each day. Location, motion and intensity of activity data were used as input to an Artificial Neural Network (ANN) model, aiming at deriving a time-activity model based solely on sensor data. The independent variables that fed the ANN input layer were consisted of a) personal temperature, *Temp*, derived from the wearable temperature sensor, b) the change of temperature with time, *dTemp/dt*, c) personal speed, derived from the GPS devices wore by the participants, *Speed,* e) the observed temperature, derived from a meteorological station located in the historical centre of Thessaloniki, $Temp_{out}$, d) and the ratio of the personal temperature to the observed one, $Temp/Temp_{out}$. Moreover, information on day light - whether it is day or night based on time - was transformed into a categorical element (day *or* night) which was also included as an input variable. The initial database was divided into training and validation set (85% and 15% of the total record entries, respectively) and the models developed from the training set were tested using the validation set.

Using a Monte Carlo analysis, distributions of participant's movement and activities - derived from the sensors experiments - were extrapolated to a larger population. The final distribution of a representative sample helped us to define the way with which people are moving in time and space (what time they start/finish work/school, their speed) as well as their different types of activity (sleeping, working, resting etc.) within the boundaries of a city. This was valuable information that was then translated into moving agents inside the GAMA platform.

By importing GIS files on the GAMA platform, we projected a simplified neighbourhood of Thessaloniki with its road (shapefile consisted of lines) and buildings (shapefile consisted of polygons) network as well as information on land use (nature of buildings: residential, office, park etc.). In this example land use information is arbitrarily defined by the modeller. The main advantage of geographical data agentification is the possibility to manage geographical objects, such as buildings, roads and people, inside the simulation by giving them a dynamic: an internal state and a behaviour. For the purposes of this paper and from now on, the term "agents" is only referring to people agents.

Agents are clustered in three age groups (children (n = 60), adults (n = 100) and elderly (n = 60)). When the model is initialized, agents are randomly distributed/allocated to a residential place which will serve as their house for the whole simulation. Following the same approach, children and adults are arbitrarily assigned to a school or an office respectively. Depending on their age, agents are programmed to move (walk on the road) from a household to either their office or their school whereas agents that belong to the elderly group stay at home. The model captures a representative day with a step of 10 minutes (one day is composed of 144 simulation cycles). Useful information on agent's routine was derived by the sensors experiments. The agents' trajectories, derived by the coded routine, are captured as points (1 point captured in every cycle) and are exported as a GIS shapefile together with a database that contains the agents' coordinates as they move in time through different locations/microenvironments. The main assumptions taken into account during the development of the ABM model are highlighted below:

- The population of the model is arbitrarily defined by the modeller. In the near future such kind of information will be based on official censuses data.
- Simplified neighborhoud of Thessaloniki: land use information is randomly defined by the modeller. In the near future, land use information will be exported by official sources so that there is an accurate representation of the urban landscape.
- Random distribution of agents – random allocation to households and random assignement to an office/school. In the near future such kind of information will be based on official census data.
- Agents choose the shortest path to reach their target (office, school, their home). In the future, the model will include spatially explicit information on habitat, roads, topography, local resources, and the effects these have on individual behavior.

The activity of an individual participant, "agent", was projected on a single layer, superposed onto urban air quality maps. These maps are based on daily average outdoor concentration data of particulate matter at various aerodynamic diameters (PM10, PM2.5 and PM1), measured and modelled for the city ofThessaloniki, taking into account traffic and non-traffic sources for the same period of time. In addition, indoor, outdoor and in-transit concentration data of particulate matter at various aerodynamic diameters were also measured using an optical counter (Grimm 11-R). This combination of modelling and measuring techinques lead us to a pool of representative values for outdoor and indoor exposure. People's exposure to air pollutants was then evaluated by assigning pollutant concentrations to a person depending on his/her coordinates as they move in time through different locations/microenvironments. The population uptake of different PM size fractions was further investigated using a human respiratory tract (HRT) deposition model.

PM levels and size distribution varied among the different parts of the urban agglomeration, as well as among different micro-environments and transportation means. Based on movement and activity, changes in exposure levels were calculated not only for the overall agent population but for individual agents as well. By estimating the daily time-activity patterns (as predicted by the coupled sensors-ANN-ABM platform) of vulnerable subgroups such as children or the elderly, we were able to estimate their personal exposure and uptake profile, taking into account a) PM size distribution across the different locations encountered and b) differences in intensity of activity.

Overall, exposure to PM and especially deposition of PM of lower fractions across the HRT are higher for children, as a result of the higher fraction time spent outdoors and in transport, the higher bodyweight-normalised inhalation rate, the more intensive activity profile and age-dependent differences in respiratory physiology.

This study represents the first step towards improving the calculation process of population exposure to environmental substances so that we would be able to draw better conclusions on the association between environment and health. Data collected by "smart" devices can help provide more accurate exposure assessment for exposure simulation modelling and epidemiology studies. Opportunities will come closer for further wide spread use to assess, at least part of the exposome. Such kind of investigations offer valuable information on the utility of several commercial devices as modular add-ons to exposure studies.

Copyright © American Institute of Chemical Engineers. All rights reserved.

435400 A Novel Technique for Determining Adsorption Breakthrough Time and Capacity Using a Non-Contact High Resolution Microwave Sensor

Abstract:

In this study, a newly developed non-contact high resolution microwave sensor was used to determine breakthrough time and adsorption capacity for different adsorbents/adsorbates with different dielectric properties. Beaded activated carbon (BAC, microwave absorbing) and a DOWEX Optipore polymeric adsorbent (V503, microwave transparent) were completely loaded with 1,2,4-trimethylbenzene (non-polar) or 2-butoxyethanol (polar). During adsorption, the changes in dielectric properties of the adsorbent were monitored in terms of two microwave parameters, quality factor and resonance frequency. Adsorption tests were completed at select relative pressures of adsorbates in the influent stream (0.03, 0.1, 0.2, 0.4, and 0.6). For all tests, the difference between the 5% breakthrough time and the time that the quality factor was 0.95 of its final value, was less than 5%. Also, for all tests, a proportional relationship between adsorption capacity and the final value of the resonance frequency shift was found. The results from this study demonstrate the potential of non-contact high resolution microwave sensors for fast and accurate measurement of the adsorption breakthrough time and adsorption capacity.

Copyright © American Institute of Chemical Engineers. All rights reserved.

434638 Studying the Effect of Flow-Rate on Bacterial Growth with Spri

Abstract:

When bacteria find the proper environmental conditions they will form irreversible attachments on the surface, attracting more bacteria, increasing their local concentration. When a certain threshold of bacteria is reached they will start to shield themselves with an exopolysaccharide (EPS) matrix forming a biofilm. Bacteria inside biofilm are known to be up to 1000 times more resistant to different antibiotics. It is essential to prevent bacterial initial attachment and further biofilm formation.

In this study, Surface Plasmon Resonance imaging (SPRi) was used to real-time monitor the effect of different flow rates on attachment and growth of *Staphylococcus aureus*. SPRi sensor is capable of label-free detection over $1cm^2$ field of view. This sensor is connected to a fluidic system allowing the control of nutrient flow rates throughout the entire experiment. Increasing the flow rate will induce the shear stress on the surface and will prevent the initial attachment of bacteria. One the other hand, higher flow rate provides more fresh food for bacteria to grow over time.

Using SPRi the effect of flow rate on *S. aureus* growth was studied over 24 hours. 15μl of *S. aureus* bacteria media were loaded on the sensor surface and fresh Lysogeny Broth (LB) media was continuously flowed over the surface at 10μl/min (slow flow rate) and 120 μl/min (fast flow rate). The internal CCD camera provides difference images of the sensor surface every three seconds which allows real-time monitoring of the entire surface as bacteria adhere and grow on it. The results obtained from comparing these images showed higher bacterial growth under lower flow rate. As a validation to visually compare bacterial growth, after each experiment the sensor surface was removed from the instrument and the surface was stained with crystal violet which stains the live bacteria. The staining results showed higher bacterial growth under lower flowrate which were consistent with the results from SPRi study.

Copyright © American Institute of Chemical Engineers. All rights reserved.

425416 Real-Time Monitoring of Bacterial Biofilm Growth and Removal Using a Quartz Crystal Microbalance

Abstract:

Bacteria are known to form complex extracellular matrices known as biofilms that allow them to adhere to multiple surfaces and thrive in diverse environments. With the growing presence of highly persistent bacterial species in the hospital setting, as recently identified by the Centers for Disease Control and Prevention, understanding biofilm growth will play a key role in combating antimicrobial resistance in modern day healthcare. Platforms and techniques for studying these bacteria as they adhere and grow on surfaces as well as strategies for their removal from surfaces can provide significant insight in preventing biofilm formation. To address these needs, a quartz crystal microbalance (QCM) was used to investigate bacterial biofilm growth and removal.

QCMs measure the precise oscillation of a quartz sensor at their resonant frequency when an alternating voltage is applied. A frequency shift response that is measured by the QCM in real time indicates either mass deposition or removal from the surface. These highly sensitive mass measurements can be used to monitor how bacteria adhere, grow, and form biofilms on surfaces.

Bacterial species *Pseudomonas aeruginosa* and *Staphylococcus aureus* were grown overnight in lysogeny broth growth media at 37 °C. Bacterial species were diluted and loaded into the QCM, where they were allowed to adhere and grow on the gold-coated quartz surface over the course of several hours. Changes in both frequency and damping were recorded to monitor bacterial accumulation on the surface. Once the bacterial cells were stably attached to the surface, an SDS detergent was loaded into the system to study the effect of biofilm removal. Biofilm removal was determined by the sensor's response as it returned to the original baseline and confirmed by visual inspection using the device's optical window. This study validates the use of a QCM for studying bacterial adhesion and removal from surfaces as well as investigating the effectiveness of biofilm removal agents.

Copyright © American Institute of Chemical Engineers. All rights reserved.

426825 Manipulation of Magnetic Beads on a Centrifugal Platform

Abstract:

In this work, the technique to manipulate the magnetic beads on the centrifugal platform is presented. Magnetic beads have been widely used in bioassays for their advantages in increasing the surface to volume ratio and improving the assay conversion by their mobility. They have been used to promote the mixing efficiency and the antibody-carriers in immunoassays on the centrifugal platform. However, an in-depth study which characterizes the interaction of magnet to magnet and magnet to magnetic beads was not available. Therefore, in this work, we are aiming to develop a working principle to manipulate the magnetic beads, which can also be used as the tool for microfluidic designs. Experiments were conducted to quantify the effective range for the magnet to magnet and magnet to magnetic beads attractions under various rotational speeds. A design principle, which was derived from the data above, was used in designing the fluidic functions for incubation and washing procedures in a disk-based immunoassays.

Copyright © American Institute of Chemical Engineers. All rights reserved.

426915 Characterization of Liquid Mobility Under Alternate Spinning

Abstract:

In this work, we present an investigation on liquid mobility under alternate spinning on a centrifugal platform. Alternate spinning, which is defined by alternating clockwise and counterclockwise rotations, is a unique function on the centrifugal platform. It has been used to perform microfluidic functions such as mixing, and decanting to assist sample preparations. In our preliminary study, we found that the mobility of the blood plasma was very different from other aqueous solutions. In order to design a working fluidic function, it is important to understand the factors which affect the liquid mobility under alternate spinning.

The liquid mobility is quantified by the elevated volume ratio of liquids with various properties. Experimental results showed that although the liquid mobility is affected by liquid viscosity, it is consistent for the liquids with viscosities between 1 to 15 cP. (The viscosity of the blood is around 2 cP) In addition, interfacial tension and contact angle do not show prominent effect on liquid mobility. Contact angle hysteresis, which is the differences between the advancing and receding contact angles, seems to play an important role on liquid mobility under alternate spinning. The relationship between the contact angle hysteresis and the liquid mobility for various liquids and surface properties was studied and a design principle to deal with various biological fluids under alternate spinning was also presented.

Copyright © American Institute of Chemical Engineers. All rights reserved.

AUTHOR INDEX

AIChE
120 Wall Street, FL 23
New York, NY 10005-4020

ISBN 978-1-5108-1846-0

CURRAN ASSOCIATES INC.
proceedings.com